Selected Titles in This Series

613 **D. L. Flannery,** The finite irreducible linear 2-groups of degree 4, 1997

612 **Joan Porti,** Torsion de Reidemeister pour les variétés hyperboliques, 1997

611 **D. Ginzburg, I. Piatetski-Shapiro, and S. Rallis,** L functions for the orthogonal group, 1997

610 **Mark Hovey, John H. Palmieri, and Neil P. Strickland,** Axiomatic stable homotopy theory, 1997

609 **Liviu I. Nicolaescu,** Generalized symplectic geometries and the index of families of elliptic problems, 1997

608 **Christina Q. He and Michel L. Lapidus,** Generalized Minkowski content, spectrum of fractal drums, fractal strings, and the Riemann zeta-functions, 1997

607 **Adele Zucchi,** Operators of class C_0 with spectra in multiply connected regions, 1997

606 **Moshé Flato, Jacques C. H. Simon, and Erik Taflin,** Asymptotic completeness, global existence and the infrared problem for the Maxwell-Dirac equations, 1997

605 **Liangqing Li,** Classification of simple C^*-algebras: Inductive limits of matrix algebras over trees, 1997

604 **Hajnal Andréka, Steven Givant, and István Németi,** Decision problems for equational theories of relation algebras, 1997

603 **Bruce N. Allison, Saeid Azam, Stephen Berman, Yun Gao, and Arturo Pianzola,** Extended affine Lie algebras and their root systems, 1997

602 **Igor Fulman,** Crossed products of von Neumann algebras by equivalence relations and their subalgebras, 1997

601 **Jack E. Graver and Mark E. Watkins,** Locally finite, planar, edge-transitive graphs, 1997

600 **Ambar Sengupta,** Gauge theory on compact surfaces, 1997

599 **Tai-Ping Liu and Yanni Zeng,** Large time behavior of solutions for general quasilinear hyperbolic-parabolic systems of conservation laws, 1997

598 **Valentina Barucci, David E. Dobbs, and Marco Fontana,** Maximality properties in numerical semigroups and applications to one-dimensional analytically irreducible local domains, 1997

597 **Ragnar-Olaf Buchweitz and John J. Millson,** CR-geometry and deformations of isolated singularities, 1997

596 **Paul S. Bourdon and Joel H. Shapiro,** Cyclic phenomena for composition operators, 1997

595 **Eldar Straume,** Compact connected Lie transformation groups on spheres with low cohomogeneity, II, 1997

594 **Solomon Friedberg and Hervé Jacquet,** The fundamental lemma for the Shalika subgroup of $GL(4)$, 1996

593 **Ajit Iqbal Singh,** Completely positive hypergroup actions, 1996

592 **P. Kirk and E. Klassen,** Analytic deformations of the spectrum of a family of Dirac operators on an odd-dimensional manifold with boundary, 1996

591 **Edward Cline, Brian Parshall, and Leonard Scott,** Stratifying endomorphism algebras, 1996

590 **Chris Jantzen,** Degenerate principal series for symplectic and odd-orthogonal groups, 1996

589 **James Damon,** Higher multiplicities and almost free divisors and complete intersections, 1996

588 **Dihua Jiang,** Degree 16 Standard L-function of $GSp(2) \times GSp(2)$, 1996

587 **Stéphane Jaffard and Yves Meyer,** Wavelet methods for pointwise regularity and local oscillations of functions, 1996

586 **Siegfried Echterhoff,** Crossed products with continuous trace, 1996

585 **Gilles Pisier,** The operator Hilbert space OH, complex interpolation and tensor norms, 1996

(Continued in the back of this publication)

The Finite Irreducible Linear 2-Groups of Degree 4

MEMOIRS
of the
American Mathematical Society

Number 613

The Finite Irreducible Linear 2-Groups of Degree 4

D. L. Flannery

September 1997 • Volume 129 • Number 613 (first of 4 numbers) • ISSN 0065-9266

American Mathematical Society
Providence, Rhode Island

1991 *Mathematics Subject Classification.*
Primary 20C15; Secondary 20D15.

Library of Congress Cataloging-in-Publication Data

Flannery, D. L. (Dane Laurence), 1965–
The finite irreducible linear 2-groups of degree 4 /D. L. Flannery.
p. cm. — (Memoirs of the American Mathematical Society, ISSN 0065-9266 ; no. 613)
"September 1997, volume 129, number 613 (first of 4 numbers)."
Includes bibliographical references.
ISBN 0-8218-0625-4
1. Representations of groups. 2. Nilpotent groups. I. Title. II. Series.
QA3.57 no. 613
[QA176]
510 s—dc21
[512′.2] 97-21341
CIP

Memoirs of the American Mathematical Society

This journal is devoted entirely to research in pure and applied mathematics.

Subscription information. The 1997 subscription begins with number 595 and consists of six mailings, each containing one or more numbers. Subscription prices for 1997 are $414 list, $331 institutional member. A late charge of 10% of the subscription price will be imposed on orders received from nonmembers after January 1 of the subscription year. Subscribers outside the United States and India must pay a postage surcharge of $30; subscribers in India must pay a postage surcharge of $43. Expedited delivery to destinations in North America $35; elsewhere $110. Each number may be ordered separately; *please specify number* when ordering an individual number. For prices and titles of recently released numbers, see the New Publications sections of the *Notices of the American Mathematical Society.*

Back number information. For back issues see the *AMS Catalog of Publications.*

Subscriptions and orders should be addressed to the American Mathematical Society, P. O. Box 5904, Boston, MA 02206-5904. *All orders must be accompanied by payment.* Other correspondence should be addressed to Box 6248, Providence, RI 02940-6248.

Memoirs of the American Mathematical Society is published bimonthly (each volume consisting usually of more than one number) by the American Mathematical Society at 201 Charles Street, Providence, RI 02904-2294. Periodicals postage paid at Providence, RI. Postmaster: Send address changes to Memoirs, American Mathematical Society, P. O. Box 6248, Providence, RI 02940-6248.

This publication is indexed in *Science Citation Index*®, *SciSearch*®, *Research Alert*®, *CompuMath Citation Index*®, *Current Contents*®/*Physical, Chemical & Earth Sciences.*
Printed in the United States of America.

♾ The paper used in this book is acid-free and falls within the guidelines established to ensure permanence and durability.
Visit the AMS homepage at URL: `http://www.ams.org/`

10 9 8 7 6 5 4 3 2 1 02 01 00 99 98 97

Contents

Abstract

We present a classification of the finite irreducible 2-subgroups of $GL(4, \mathbb{C})$; that is, we give a parametrised list of representatives for the conjugacy classes of such groups.

Each group listed is defined by a generating set of monomial matrices. There are essentially three possibilities for the projection of an irreducible monomial 2-group into the group of all permutation matrices. The classification problem accordingly falls into three separate cases. Each case may be handled by a general method consisting of three major steps. Techniques for applying the method to the most difficult case are developed in detail, so that the other two cases may then be dealt with routinely. The techniques used include elementary character theory, a method for drawing the Hasse diagram of the submodule lattice of a direct sum, and cohomology theory, particularly the calculation of 2-cohomology by means of the Lyndon-Hochschild-Serre spectral sequence.

Related questions concerning isomorphism between the listed groups, and Schur indices over $\mathbb{Q}$, are also considered.

1991 *Mathematics Subject Classification.* Primary 20C15, Secondary 20D15.

Introduction

By *a linear group of degree* n, we mean a subgroup of $GL(n) = GL(n, \mathbb{C})$. Each class of linear groups that we consider is a union of conjugacy classes. To "list" the linear groups in a class means to provide a complete list of representatives of their $GL(n)$-conjugacy classes.

Let p be a prime. The problem of listing the finite irreducible linear p-groups of degree n (necessarily a power of p) is considered for $n = p$ by Conlon in [5]. The listing problem for $n = p^2$ and general p is extremely difficult. Our purpose in this monograph is to present a solution of the listing problem in the special case $n = 4$. Even in this smallest possible prime-square degree, substantial difficulties exist.

The list we obtain, like Conlon's, is infinite. In both lists, the conjugacy class representatives which appear are sorted into a finite number of families. The members of any one family admit a common description in terms of integer parameters: each member of a family is labelled by a unique parameter string. This description amounts to giving a generating set of monomial matrices for each group. The entries of these matrices are explicit functions of the parameters.

One motivation for studying this type of listing problem is the desire in computational group theory for better soluble quotient algorithms. Typically, these algorithms accept as input a finite presentation of a group, and produce as output a list of all "sufficiently small" finite soluble quotients of the group with that presentation. Some such algorithms envisaged will employ databases containing all sufficiently small finite soluble groups which have faithful primitive permutation representations, each necessarily of prime power degree. Listing groups of this kind, in degree p^m, is equivalent to listing irreducible soluble subgroups of $GL(m, p)$. Irreducible soluble subgroups of $GL(m, p)$ are closely related—for example, are abstractly isomorphic—to subgroups of $GL(m)$. Consequently, the list given in this monograph serves as a preparatory step toward extending presently available databases. The next step in that direction is to obtain lists of irreducible 2-subgroups of $GL(4, p)$ for each odd prime p.

Received by the editor March 11, 1996.

Overview

The monograph is organised into seven chapters. Chapter 1 is devoted to setting out background concerned with module theory, submodule lattices and linear group theory that is pertinent to the main problem. Although our primary aim is construction of conjugacy class representatives, it is worthwhile understanding isomorphism between the listed groups. Specifically, in Chapter 2 we ask whether the concepts of linear isomorphism (conjugacy) and abstract isomorphism are equivalent, for finite irreducible 2-subgroups of $GL(4)$. When the groups satisfy a certain uniqueness condition, this question may be answered using the ideas from cohomology theory set down in Chapter 2 and with reference to the full list of groups given in Chapter 6.

In order to survey how the remaining material is organised, we will now examine the basic features of the main problem in more detail.

Elements of the n-dimensional $\mathbb{C}$-space $\mathbb{C}^{(n)}$ will be considered as column vectors, with $GL(n)$ acting on the left. The group of all permutation matrices in $GL(n)$ acts as the symmetric group of degree n on the ordered basis $\{e_i \mid 1 \leq i \leq n\}$ of $\mathbb{C}^{(n)}$, where e_i has ith entry 1 and jth entry 0, $j \neq i$. So the same notation S_n will be used for both groups. The group $M(n)$ of monomial matrices in $GL(n)$ is the permutational wreath product $\mathbb{C}^\times \operatorname{wr} S_n$, where $\mathbb{C}^\times$ denotes the multiplicative group of $\mathbb{C}$. The base group of this wreath product is the group $D(n)$ of all diagonal matrices in $GL(n)$.

A finite p-subgroup of $GL(n)$ is conjugate to a subgroup of $M(n)$. Consider the canonical projection of $M(n)$ onto S_n. The restriction of this homomorphism to a subgroup G of $M(n)$ has kernel $D(n) \cap G$, which we call the *diagonal subgroup of* G, and image $T = D(n)G \cap S_n$, which we call the *projection group of* G. Stating that G has projection group T is equivalent to stating that $D(n)G = D(n)T$.

Proposition 1 *Let B denote the Sylow p-subgroup of $D(n)$; that is, the direct product of n copies of the quasicyclic p-group C_{p^∞}. A finite p-subgroup G of $GL(n)$ is conjugate to a subgroup of BP, where P is a Sylow p-subgroup of S_n. Furthermore, if G is a finite irreducible p-subgroup of BS_n then $BG = BT$ for some transitive p-subgroup T of S_n.*

Proof. The first assertion follows from Theorems II.4 and IV.2 of [14]. Assume G as in the second assertion, with projection group T. If T were intransitive then a proper nonempty subset of the basis $\{e_i \mid 1 \leq i \leq n\}$ of $\mathbb{C}^{(n)}$ would be invariant under T. Then a proper non-trivial subspace of $\mathbb{C}^{(n)}$ would be invariant under BT and hence under G, contradicting the irreducibility of G. □

Thus, before we can list the finite irreducible linear p-groups of degree p^m, we must first list the transitive p-subgroups of S_{p^m}. These are well-known for $m \leq 2$ and have been studied for $m = 3$ in [2]. For larger values of m this initial classification problem is challenging in its own right.

From now on, B in Proposition 1 is defined for the special case $p = 2$ and $n = 4$. A transitive 2-subgroup T of S_4 either has order 4, or is dihedral of order 8. The cyclic subgroups of S_4 of order 4 are transitive, and form a single conjugacy class in S_4. The same is true of the dihedral subgroups of order 8 (Sylow 2-subgroups). We choose a particular subgroup from each of these conjugacy classes by setting $a = (1,2)(3,4)$, $c = (1,2,3,4)$, and then defining $C = \langle c \rangle$, $D = \langle a, c \rangle$. The only transitive fours group in S_4 is the normal subgroup $V_4 = \langle a, b \rangle$, where $b = (1,3)(2,4) = c^2$. We see that a finite irreducible 2-subgroup of BS_4 is conjugate (by a permutation matrix) to one whose projection group is contained in D. By Proposition 1, the main problem is therefore reduced to:

> for each case $T = V_4$, $T = C$ and $T = D$, compile a list of all finite irreducible subgroups G of BT such that $BG = BT$, ensuring that a group in any one list is not conjugate to a group in any other.

The most difficult part of this reduced problem is the listing for $T = V_4$; it is dealt with in Chapter 3. The remaining parts of the problem are then dealt with briefly in Chapters 4 and 5, for the most part by mimicking the techniques used in Chapter 3. A discussion of these techniques now follows.

For each T, B is a T-module (formally, $\mathbb{Z}T$-module). Suppose that $G \leq BS_4$ and $BG = BT$: that is, G is an extension of the finite T-submodule $B \cap G$ of B by the projection group T of G. The first major task that we undertake is the construction of finite T-submodules of B, thereby determining all possibilities for $B \cap G$. These submodules fall into finitely many infinite families, the multiplicity of which contributes significantly to the complex nature of the main listing problem. Subsequent argument depends upon having information about the inclusion relations between the submodules. (Additionally, such information would assist in translating the relevant part of the full list of linear groups over $\mathbb{C}$ into lists of irreducible 2-subgroups of $GL(4,p)$ for odd primes p.) The object of interest in this regard is therefore the poset, or lattice, of finite T-submodules of B. A poset is described by means of its Hasse diagram; when the poset is the lattice of finite submodules of a module, an edge in its associated Hasse diagram represents maximal inclusion, and knowledge of these edges enables one to decide whether or not two given finite submodules are comparable. We show that a critical part of the lattice of finite T-submodules of B is essentially the submodule lattice of a direct sum. Methods for drawing the Hasse diagram of such a submodule lattice are given in [10]. These will be applied in Section 3.1 for the case $T = V_4$.

Of course, G is not necessarily determined by its diagonal subgroup and projection group. Given a transitive 2-subgroup T of S_4 and an arbitrary finite T-submodule U of B, we need to solve the "extension problem" of finding (up to a suitable equivalence) all finite subgroups G of BT such that $B \cap G = U$ and $BG = BT$. (Although such a group G is not necessarily irreducible, it is easy to recognise the reducible G.) We begin our solution of this problem by establishing a general method for calculating the order of the second cohomology group of T with coefficients in U. When T is not cyclic, this calculation is based on the construction of an appropriate spectral sequence converging to the cohomology of T. (When T is cyclic, a standard and more direct approach is possible.) The orders so calculated, in conjunction with a result proved in Chapter 2, may then be used in listing representatives of the BT-conjugacy classes of finite irreducible subgroups of BT with projection group T. However, groups in the list obtained may still be conjugate in $GL(4)$, and the "conjugacy problem" is to determine all remaining conjugacies. These procedures are carried out for $T = V_4$ in Sections 3.2 and 3.3.

In Chapters 4 and 5 we apply variants of the techniques discussed above to the cases $T = C$ and $T = D$, respectively. We also determine all instances of conjugacy and isomorphism between groups in any two of the three derived lists.

At this stage we will have completely solved the main listing problem. A full list of the finite irreducible 2-subgroups of $GL(4)$ is given in Section 6.1. We will also be in a position to answer a slightly weaker version of the isomorphism question from Chapter 2, and do so in Section 6.2.

As a matter of interest, in the concluding Chapter 7 we show how to calculate the Schur index over $\mathbb{Q}$ for characters of each group in the list of Section 6.1.

Notation

The notation B, C, D, V_4 and a, b, c introduced above is fixed throughout the monograph.

Group theory notation is standard and mostly follows [17]. The centre of a group G is denoted $Z(G)$, and its Frattini subgroup is denoted $\Phi(G)$. The commutator $g^{-1}h^{-1}gh$ of two elements g and h of G is written $[g, h]$. If G is a p-group then $\Omega_k(G)$ is the subgroup generated by all elements whose orders divide p^k (so that if G is an abelian p-group, then $\Omega_k(G)$ is the subgroup consisting of all elements whose orders divide p^k).

Maps and actions are mostly written on the right, although there are exceptions, such as representations and action by linear groups.

Each module discussed is defined over an associative ring R with 1. By the exponent of an R-module M, we mean its exponent as an (additive) abelian group. If $\mathcal{S}$ is a subset of M then $\langle \mathcal{S} \rangle$ denotes the subgroup generated by $\mathcal{S}$.

An element of $D(n)$ is written as the ordered n-tuple of its main diagonal entries. Elements of S_4 are considered both as abstract permutations and as permutation matrices, and the same notation is used in each case. For example, $(1,2)$ denotes the element of $GL(4)$ that results from interchanging the first and second rows of the identity matrix.

The terminology and notation of [10] relating to submodule lattices and their Hasse diagrams are carried over. For instance, a vertex in the Hasse diagram of the submodule lattice of a direct sum of two modules is classified as either *Cartesian* or *non-Cartesian*; an edge is classified as one (and only one) of the types *Cartesian*, *restriction* or *composition.* (In [10], the non-Cartesian submodules are called *diagonal*, but we will not use that designation here.) These classifications of vertices and edges depend on the choice of direct decomposition. If M is an R-module then $\mathcal{L}(M)$ denotes the lattice of finite submodules of M, and $\mathcal{H}(M)$ denotes the Hasse diagram of $\mathcal{L}(M)$. The vertex set and edge set of $\mathcal{H}(M)$ are denoted $\mathcal{V}(M)$ and $\mathcal{E}(M)$, respectively.

The vector space of dimension n over the field of q elements is denoted $V(n,q)$, and the subspace lattice of $V(n,q)$ is denoted $\mathcal{L}(n,q)$.

Unless stated otherwise, group characters are complex.

Acknowledgements

This work is an expansion of the research carried out for my PhD thesis [9], and is formally dedicated to my PhD supervisor, Dr L.G. Kovács. I am immensely grateful to Dr Kovács for his guidance, patient teaching, and constant generosity. His influence on this work is profound; it is unlikely that I would have been drawn to some of the ideas and results contained herein without his help.

I thank Dr W. Feit for directing my attention to the problem considered in Chapter 7 and for suggesting the reductive approach to its solution.

I acknowledge comments made by a referee for the Journal of Algebra, who read an earlier paper of mine on a different way of obtaining the results in the first part of Section 3.2 of this monograph (Propositions 3.2.1–3.2.3). The referee pointed out that the method of calculating with spectral sequences used in Section 3.2 is more natural and efficient than the one used in the earlier paper.

Chapter 1

Preliminaries

1.1 Module theory

Let T be a regular subgroup of S_m, $m \geq 1$. The finite normal subgroups in the base group of $C_{p^\infty} \operatorname{wr} T$ may be viewed in a group algebraic context, as follows. If U is the largest finite normal subgroup with exponent p^n, then the regularity of T allows one to define an obvious isomorphism of U onto A_A, where A is the group algebra $(\mathbb{Z}/p^n\mathbb{Z})T$. This isomorphism sets up an inclusion-preserving bijection between the set of finite normal subgroups with exponent at most p^n in the base group of $C_{p^\infty} \operatorname{wr} T$, and the set of submodules of A_A. In the sequel, when translating between these two contexts, the relevant isomorphism and bijection will be implicitly assumed.

To list the finite T-submodules of B for each transitive 2-subgroup T of S_4, it is clearly sufficient to deal with the cases $T = V_4$ and $T = C$ only. Since T is regular in both cases, the equivalence noted above may be exploited in the listing of submodules. We do so in Section 3.1 and Chapter 4. Some ring and module theory required for that purpose is recorded below.

Many of the nonzero modules that we deal with are finitely generated and hence have maximal submodules. For the next result, see 5.25, p.115 of [7].

Proposition 1.1.1 *Let R be a commutative local ring such that $R/\operatorname{rad}R$ is a field of characteristic p, and let H be a finite p-group. Then $RH/\operatorname{rad}RH \cong R/\operatorname{rad}R$, so that RH is a local ring and there is a unique isomorphism type of simple RH-module.*

In particular, if $R = \mathbb{Z}/p^n\mathbb{Z}$ then each simple RH-module has order p.

Proposition 1.1.2 *Let R be a self-injective commutative local ring and let H be a finite abelian group. Then two submodules of RH_{RH} are isomorphic if and only if they are equal.*

Proof. The hypotheses imply that RH is a self-injective ring (see Exercise 2(d), p.402 of [6]), and so a homomorphism between submodules of RH_{RH} may be extended to an endomorphism of RH_{RH}. Such an endomorphism is multiplication on the left by an element of RH, viz. the image of 1 under the endomorphism. Commutativity of RH then gives the conclusion. □

Since $\mathbb{Z}/p^n\mathbb{Z}$ is self-injective, Proposition 1.1.2 implies that two finite T-submodules of B are isomorphic if and only if they are equal. Further consequences of Propositions 1.1.1 and 1.1.2 are that A_A has a unique simple submodule, soc$A$, and a unique maximal submodule, rad$A$. As subgroups of A, these have generators as specified in the next proposition.

Proposition 1.1.3 *For $n \geq 1$, set $R = \mathbb{Z}/p^n\mathbb{Z}$ and let H be a finite p-group. Then* $\text{rad}RH = \langle p,\ x-1 \mid x \in H, x \neq 1\rangle$ *and* $\text{soc}RH = \langle p^{n-1}\sum_{x\in H} x\rangle$.

Every finite non-trivial T-submodule of B contains the unique simple T-submodule of B. This submodule, which we denote B_0, is generated by $x_0 = (-1,-1,-1,-1)$.

We close this section with some comments about duality. As usual, if U is any (right) R-module, then $\text{ann}_R U$ denotes the (right) annihilator of U in R. If R and H are as in Proposition 1.1.2 then there exists a non-degenerate associative symmetric bilinear form $f: RH \times RH \to R$. Fix such a form f. If U is a submodule of RH_{RH} then the dual $U^\perp$ of U is $\{y \in RH \mid f(x,y) = 0 \text{ for all } x \in U\} = \text{ann}_{RH} U$.

Proposition 1.1.4 *Let R be a finite self-injective commutative ring such that $R/radR$ is a field of characteristic p, and let H be a finite p-group. If U is a submodule of RH_{RH} then*

(i) $(U^\perp)^\perp = U$

(ii) $|U||U^\perp| = |RH|$.

Furthermore, if U is singly generated then

(iii) $RH_{RH}/U^\perp \cong U$.

Proof. Property (i) is the so-called "double annihilator property"—see Exercise 13, p.286 of [1]. An easy induction on $|U|$ gives (ii). If $U = uRH$ then the homomorphism $x \mapsto ux$ of RH_{RH} onto U has kernel $\text{ann}_{RH}U$, yielding (iii). □

1.2 Submodule lattices and Hasse diagrams

The lattice of finite normal subgroups in the base group of C_{2^∞} wr V_4 has an associated directed graph, or Hasse diagram, defined in the usual way. A preliminary step in the program outlined in the Introduction is to obtain a description of this (infinite) graph. As we will show, this problem can be cast as a special case of the general problem considered at length in [10]: describing the Hasse diagram of the submodule lattice of a direct sum of two modules. A method for determining the vertices of this Hasse diagram is given in Section 3 of [10]. To describe the edges, we employ the concepts of *page* (a distinguished sublattice) and *atlas* (the collection of all pages), discussed in Section 5 of [10]. These concepts are defined relative to the submodule lattice and its Hasse diagram. An atlas is *complete*, meaning that each edge in the full Hasse diagram occurs in the Hasse diagram of at least one of the atlas pages—so that if we are dealing with finite submodules only, then the vertices of the full Hasse diagram are likewise covered by the Hasse diagrams of the atlas pages. For efficiency, we may eliminate redundancy, if it occurs, by deleting some atlas pages while preserving completeness overall (this comment applies particularly to the atlas constructed in Section 3.1).

In this section we describe the Hasse diagram of the lattice of finite normal subgroups in the base group of $C_{2^\infty} \operatorname{wr} C_2$. This serves as an instructive model for the more complicated problem involving $C_{2^\infty} \operatorname{wr} V_4$, to be handled in Section 3.1. Moreover, structural information about the Hasse diagram studied in this section is a vital ingredient of the method used in Section 3.1.

Set $A = (\mathbb{Z}/p^n\mathbb{Z})C_p$, $n \geq 1$, where $C_p = \langle x \mid x^p = 1 \rangle$. By the opening remarks in Section 1.1, a finite normal subgroup in the base group of $C_{p^\infty} \operatorname{wr} C_p$ may be viewed as a submodule of A_A, for large enough n. So we focus first on $\mathcal{H}(A_A)$. (Note that each edge in this Hasse diagram stands for a module of order p.) We let p be an arbitrary prime, as this entails no more effort than restricting to the case $p = 2$ of immediate concern. Our approach may be compared with the one taken in Section 1 of [5].

By the comments made after Proposition 1.1.2, we need to describe $\mathcal{H}(\mathrm{rad}A/\mathrm{soc}A)$. We do this by writing $\mathrm{rad}A/\mathrm{soc}A$ as a direct sum and then applying the results of [10]. Denote by Z the submodule $\langle \hat{x} \rangle$ of A_A, where $\hat{x} = \sum_{i=0}^{p-1} x^i$, and by W the augmentation ideal $\langle x^i - 1 \mid 1 \leq i \leq p-1 \rangle$ of A. By Proposition 1.1.3, we see that

$$\mathrm{rad}A/\mathrm{soc}A = Z/\mathrm{soc}A \oplus W/\mathrm{soc}A. \tag{1.1}$$

Obviously $Z/\mathrm{soc}A$ is a uniserial A-module of composition length $n-1$. The action of x on $Z/\mathrm{soc}A$ is trivial. The A-module structure of $W/\mathrm{soc}A$ is summarised in the following lemma.

Lemma 1.2.1 *W is a uniserial A-module of composition length $(p-1)n$, with composition terms W_i defined inductively by $W_0 = \{0\}$ and $W_i = \langle w_i, W_{i-1} \rangle$ for $1 \leq i \leq (p-1)n$, where $w_i = (1-x)^{(p-1)n-i+1}$. Also, the only nonzero sections of W upon which x acts trivially are its composition factors W_i/W_{i-1}.*

Proof. Of course, $\mathrm{rad}W = (\mathrm{rad}A)W$. Induction establishes that $\mathrm{rad}^i W = W^{i+1}$ and $W^i/W^{i+1} = \langle (1-x)^i + W^{i+1} \rangle$, as an additive abelian group, is cyclic of order p.

Therefore, the Loewy factors of W are simple, and W is uniserial with composition length $\log_p|A| - \log_p|Z| = pn - n$. The last assertion is clear. □

The next result gives explicitly the submodules of A_A.

Proposition 1.2.2 *Define $z_i = p^{n-i}\hat{x}$ and let w_i and W_i be as defined in Lemma* 1.2.1. *Further define the following submodules of A_A:*

$$A(i,j) = \langle z_i, W_j \rangle$$

for $1 \leq i \leq n$ and $1 \leq j \leq (p-1)n$, and

$$A(i,j,k) = \langle z_{i+1} + kw_{j+1}, A(i,j) \rangle$$

for $1 \leq i \leq n-1$, $1 \leq j \leq (p-1)n - 1$ and $1 \leq k \leq p-1$. Then a proper nonzero submodule of A_A is one of the $A(i,j)$ or $A(i,j,k)$, for unique values of i,j,k.

Proof. By Theorem 3.1 of [10], the submodules of $\mathrm{rad}A/\mathrm{soc}A$ are in one-to-one correspondence with isomorphisms between sections of the direct summands in (1.1). Since the action of x on Z is trivial, isomorphic sections are either zero or simple by Lemma 1.2.1. Isomorphisms between zero sections give rise to the Cartesian submodules $A(i,j)/\mathrm{soc}A$ (with respect to (1.1)). Isomorphisms between simple sections give rise to the diagonal submodules $A(i,j,k)/\mathrm{soc}A$: for fixed i and j, each value of k specifies one of the $p-1$ isomorphisms between $\langle z_{i+1}\rangle/\langle z_i\rangle$ and W_{j+1}/W_j. □

We follow the procedure given in Section 5 of [10] to describe $\mathcal{E}(\mathrm{rad}A/\mathrm{soc}A)$. Since Z and W are both uniserial, these edges are either Cartesian or restriction—no composition edges occur. This means that an atlas of $\mathcal{L}(\mathrm{rad}A/\mathrm{soc}A)$ does not contain pages of the kind prescribed by Theorem 4.3 (v) and (vi) of [10]. Indeed, there are only two types of pages: there is the Cartesian product page $\mathcal{L}(Z/\mathrm{soc}A) \times \mathcal{L}(W/\mathrm{soc}A)$, and any other page is of the form $\mathcal{L}(A(i+1,j+1)/A(i,j))$. (In the terminology of [10], the latter page is the "interval of definition" of the non-Cartesian vertex $A(i,j,k)$ of $\mathcal{H}(\mathrm{rad}A/\mathrm{soc}A)$.) The associated Hasse diagrams are illustrated in Figure 1.1, for $p = 2$.

We are now able to describe the Hasse diagram of the lattice $\mathcal{L}$ of non-trivial finite normal subgroups in the base group of $C_{2^\infty} \,\mathrm{wr}\, C_2$. Denote this base group by K. In $GL(2)$, K appears as the group of diagonal matrices of 2-power order. In this setting, the subgroup generators z_i and w_i of the normal subgroups in K may be defined as follows. For $k \geq 0$, let $\omega_k = \exp(2^{-k}\pi\sqrt{-1})$, so that $\omega_0 = -1$, $\omega_{k+1}^2 = \omega_k$ and $\langle \omega_0, \omega_1, \ldots \rangle$ is the Sylow 2-subgroup C_{2^∞} of $\mathbb{C}^\times$. Then $z_i = (\omega_{i-1}, \omega_{i-1})$ and $w_i = (\omega_{i-1}, \omega_{i-1}^{-1})$. For $p = 2$, W as defined before Lemma 1.2.1 is inverted elementwise by x and is (additively) a cyclic group of order 2^n. Interpreting Proposition 1.2.2 in the present context, we see that the non-trivial finite normal subgroups of $C_{2^\infty} \operatorname{wr} C_2$ lying in K are

$$A(i,j) = \langle z_i, w_j \rangle, \qquad A(i,j,1) = \langle z_{i+1}w_{j+1}, w_j \rangle; \qquad i,j \geq 1.$$

We note further that $K = ZW$, where $Z = \langle z_1, z_2, \ldots \rangle$ is the subgroup of scalars in K, $W = K \cap SL(2) = \langle w_1, w_2, \ldots \rangle$, and $Z \cap W = \mathrm{soc}K = \langle z_1 \rangle$. The atlas of $\mathcal{L}$ consists of the Cartesian product page and pages of the form $\mathcal{L}(A(i+1,j+1)/A(i,j))$; their Hasse diagrams are depicted in Figures 1.1 and 1.2.

The arguments presented in this section will be extended in Section 3.1 to treat the lattice of finite normal subgroups of $C_{2^\infty} \operatorname{wr} V_4$ lying in the base group. In that example, it is possible to choose a direct decomposition which satisfies the "feasibility conditions", mentioned in Section 6 of [10], which make the atlas of the submodule lattice useful in drawing its Hasse diagram. (One such condition, also satisfied by the decomposition $K/\mathrm{soc}K = Z/\mathrm{soc}K \oplus W/\mathrm{soc}K$ considered here, is that a section in one summand isomorphic to a section in the other has comparatively small composition length.)

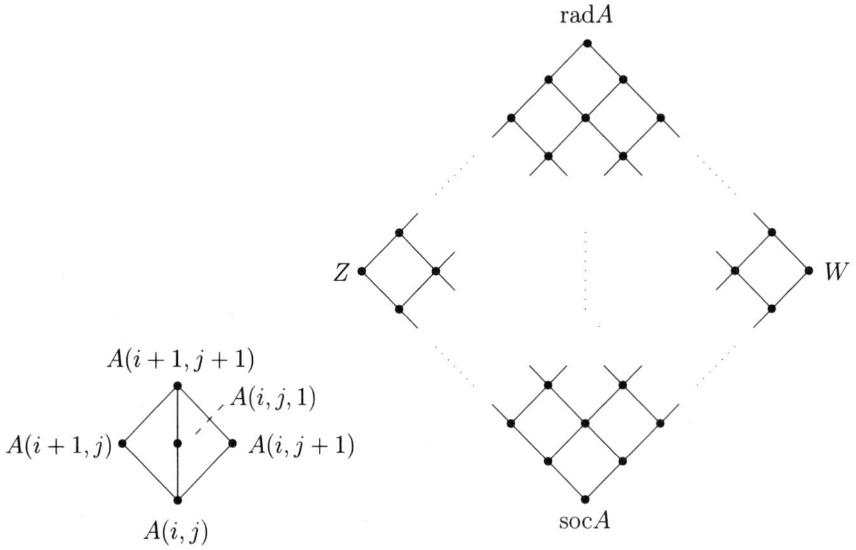

$\mathcal{L}(A(i+1,j+1)/A(i,j))$ $\qquad$ $\mathcal{L}(Z/\mathrm{soc}A) \times \mathcal{L}(W/\mathrm{soc}A)$

Figure 1.1: Hasse diagrams of pages in the atlas of $\mathcal{L}(\mathrm{rad}A/\mathrm{soc}A)$, $p = 2$

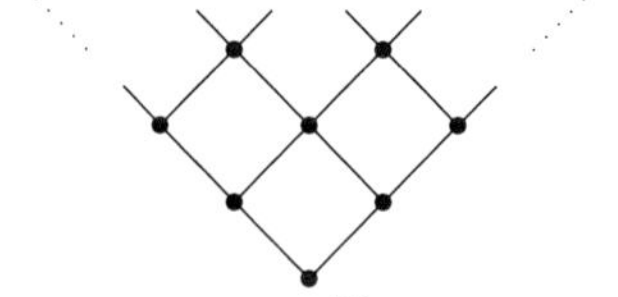

Figure 1.2: Hasse diagram of the Cartesian product page in the atlas of $\mathcal{L}$

1.3 Irreducibility

We collect together in this section some basic results about irreducibility in $GL(4)$. Chief among these is a characterisation that allows irreducibility to be recognised solely by inspection of diagonal subgroup. This characterisation makes use of the following definition: a T-module U is said to be *faithful* if $\mathsf{C}_T(U) = 1$.

The first result may be proved using elementary character theory, such as that found in Chapter 2 of [13].

Lemma 1.3.1 *Let G be a finite 2-subgroup of $GL(4)$. Then any one of the following implies that G is reducible:*

(i) $|G| \leq 16$;

(ii) *G has an abelian maximal subgroup;*

(iii) *G has an abelian normal subgroup A of index 4 such that $A < \mathsf{C}_G(A)$.*

Definition 1.3.2 Let U be a subgroup of B, and denote the (i,j)th entry of $x \in GL(4)$ by $x_{i,j}$. Then for each i, $1 \leq i \leq 4$, we define the linear character α_i of U by $\alpha_i : u \in U \mapsto u_{i,i}$.

In preparation for proof of the next two results, we now recall some facts about regular permutation groups. Let T be a regular subgroup of S_n. An automorphism of T is realised as the restriction of an inner automorphism of S_n, so that $\mathsf{N}_{S_n}(T)/\mathsf{C}_{S_n}(T) \cong \mathrm{Aut}(T)$. Also, it is easy to see that if U is a faithful T-module then $\mathsf{C}_{S_n}(U) \cap \mathsf{N}_{S_n}(T) = 1$. If T is abelian then T is its own centraliser in S_n. In particular, if $T \leq S_4$ has order 4 then $\mathsf{N}_{S_4}(T)$ splits over T: by $\langle (1,2),(2,3) \rangle$ if $T = V_4$, and by $\langle a \rangle$ if $T = C$.

Lemma 1.3.3 *Let T be a transitive 2-subgroup of S_4 and suppose that U is a faithful finite T-submodule of B. Then $\alpha_i = \alpha_j$ if and only if $i = j$.*

Proof. Each D_8 in S_4 is the normaliser of a transitive C_4, and so we may assume that $|T| = 4$. Furthermore, since T permutes transitively the diagonal entries of an element of B, it is enough to prove that $\alpha_j \neq \alpha_1$ for $j \neq 1$. Suppose that $\alpha_j = \alpha_1$ and denote the elements of $\{2,3,4\}\backslash\{j\}$ by k and l. Then there exists $u \in U$ such that $\alpha_k(u) \neq \alpha_l(u)$, for otherwise we would have $(1,j)(k,l) \in \mathsf{C}_{S_4}(U) \cap V_4 \leq \mathsf{C}_{S_4}(U) \cap \mathsf{N}_{S_4}(T) = 1$. Also, there exists $t \in T$ such that $(k^t, l^t) = (1,j)$ or $(k^t, l^t) = (j,1)$. But then $\alpha_1(u^t) \neq \alpha_j(u^t)$, a contradiction. □

Proposition 1.3.4 *Let G be a finite 2-subgroup of BS_4 such that $BG = BT$ for some transitive 2-subgroup T of S_4. If $B \cap G$ is faithful as a T-module then G is irreducible.*

Proof. By transitivity of T, the linear characters α_i of the normal subgroup $B \cap G$ of BT are pairwise BT-conjugate. Since $BT = BG$ and B is abelian, the α_i are in fact pairwise G-conjugate.

First assume that $|T| = 4$. Let α be the trace map on G and β an irreducible constituent. Since $\alpha_{B\cap G} = \sum_{i=1}^4 \alpha_i$, some α_k is an irreducible constituent of $\beta_{B\cap G}$. By the first paragraph and Lemma 1.3.3, the distinct G-conjugates of α_k are all of the α_i. Then it follows from Clifford's Theorem that $\beta_{B\cap G} = e\alpha_{B\cap G}$ for some integer $e \geq 1$. In fact $e = 1$, since the degree of $\beta_{B\cap G}$ is at most 4. Hence $\alpha = \beta$ is irreducible.

If $|T| = 8$ then clearly G has a subgroup H such that $BH = BV_4$ and $B \cap H$ is a faithful V_4-module. By the above H is irreducible, and thus so too is G. □

By Lemma 1.3.1 (iii) and Proposition 1.3.4, we have the following theorem.

Theorem 1.3.5 *Let G be a finite 2-subgroup of BS_4 such that $BG = BT$ for some transitive 2-subgroup T of S_4 of order 4. Then G is irreducible if and only if $B \cap G$ is a faithful T-module.*

By Theorem 1.3.5, when $T = V_4$ or $T = C$, we may decide irreducibility before solving the relevant extension problem. Although Theorem 1.3.5 fails when $T = D$, we will see that in this case also it is sufficient to solve the extension problem only for faithful D-submodules of B.

The following two consequences of irreducibility are of fundamental importance in our solution of the conjugacy problem.

Proposition 1.3.6 *Let G be a finite irreducible 2-subgroup of $GL(4)$ with an abelian normal subgroup A of index 4. Then there is $x \in GL(4)$ such that $G^x \leq BS_4$ and $A^x = B \cap G^x$.*

Proof. The hypotheses imply that A is diagonalisable, and so we choose $y \in GL(4)$ such that $A^y \leq B \cap G^y$. Then $A^y = B \cap G^y$ by Lemma 1.3.1 (ii). Hence, for the rest of the proof we will assume that $A = B \cap G$.

Let α be the trace map on G and let α_i be as in Definition 1.3.2. By Frobenius reciprocity, the multiplicity of the irreducible character α of G in the induced character α_i^G is equal to the multiplicity of α_i in α_A, which is 1. But α_i^G has degree $|G : A| = 4$, and so $\alpha = \alpha_i^G$ for all i. Therefore, the representations affording α and α_i^G are equivalent. The former representation is just inclusion in $GL(4)$, whereas the latter is monomial and maps $A = B \cap G$ into B. This means that there is some $x \in GL(4)$ with $G^x \leq BS_4$ and $A^x \leq B \cap G^x$; that is, $A^x = B \cap G^x$, as required. □

Proposition 1.3.7 *Let G be a finite 2-subgroup of BS_4 such that $BG = BT$ for some transitive 2-subgroup T of S_4. If $B \cap G$ is a faithful T-module then*

$$\mathsf{N}_{GL(4)}(B \cap G) \leq M(4).$$

Proof. As a $(B \cap G)$-module, $\mathbb{C}^{(4)} = \oplus_{j=1}^4 V_j$, where V_j is the subspace spanned by e_j. Choose $x \in \mathsf{N}_{GL(4)}(B \cap G)$. The $(B \cap G)$-module xV_j affords a 1-dimensional representation ϕ_j of $B \cap G$, defined by $\phi_j(g)xe_j = gxe_j$ for each $g \in B \cap G$. That is, $\phi_j(g)x_{ij} = \alpha_i(g)x_{ij}$, $1 \leq i \leq 4$. By Lemma 1.3.3, this means that each column of x has a single nonzero entry. Invertibility of x then implies that x is monomial. □

Remark 1.3.8 Suppose that G and H are finite irreducible 2-subgroups of BS_4 with $G^x = H$ for some $x \in M(4)$. Equivalently, $G^d \leq BS_4$ for some $d \in D(4)$, whence $[d, t] \in BS_4 \cap D(4) = B$ for each $t \in T = BG \cap S_4$. By transitivity of T, we may then write each entry of d as a C_{2^∞}-multiple of a fixed entry of d. Hence x acts as conjugation by an element of BS_4.

Chapter 2

The isomorphism question

Let G and H be finite irreducible subgroups of $GL(n)$. If $G \cong H$, are G and H necessarily conjugate? For $n = p$ and G, H finite p-groups, this isomorphism question is answered affirmatively, by Proposition 4.2 of [5]. For $n = 4$, we consider a restricted version of the question; namely, is an isomorphism between finite irreducible 2-groups in BS_4 mapping diagonal subgroup onto diagonal subgroup always realised as conjugacy in $GL(4)$? This situation is significant: a group in the final list that has more than one abelian normal subgroup with the same quotient as its diagonal subgroup is exceptional, and easily recognised. The restricted isomorphism question is answered affirmatively, using machinery set up in this chapter (although the actual verification is postponed until Chapter 6).

The methods used to answer the isomorphism questions here and in [5] depend ultimately on solution of the relevant full listing problem. It is evident, however, that *a priori* answers would reduce the difficulty of solving the full listing problem. In any event, our approach to the isomorphism question incorporates theoretical results that are extremely useful in solving the extension and conjugacy subproblems.

Knowledge of low-dimensional cohomology theory of finite groups will be assumed: for example, see Chapter 11 of [17] or Chapter VI of [12]. However, we review explicitly some elementary 2-cohomology theory that will be used in making later definitions.

Let U be an abelian group, written additively, and let Q be a multiplicatively written group acting on U by means of the coupling $\chi \in \mathrm{Hom}(Q, \mathrm{Aut}(U))$. Each extension $0 \to U \xrightarrow{\iota} G \xrightarrow{\pi} Q \to 1$ of U by Q gives rise to a (unique) coupling via conjugation in G and choice of a normalised transversal function $\sigma\colon Q \to G$ for π. When the coupling of an arbitrary extension of U by Q is not mentioned, it is assumed to be χ. There is a standard equivalence relation on the set of all extensions of U by Q with coupling χ. As usual, denote the abelian group of 2-cocycles from Q to U and the subgroup of 2-coboundaries by $Z^2(Q,U)$ and $B^2(Q,U)$, respectively. For brevity, $\psi B^2(Q,U) \in Z^2(Q,U)/B^2(Q,U)$ is written as $[\psi]$. With the chosen extension we associate a particular element ψ of $Z^2(Q,U)$, defined by $(x,y)\psi\iota = ((xy)\sigma)^{-1}x\sigma y\sigma$ for $x, y \in Q$ and σ as above. The map sending the equivalence class of the extension to $[\psi]$ is well-defined and independent of the choice of σ. Conversely, with each element ψ of $Z^2(Q,U)$ we associate an extension $0 \to U \to E_\psi \to Q \to 1$ with coupling χ. Here, $E_\psi = \{(x,u) \mid x \in Q, u \in U\}$ and multiplication in E_ψ is given by

$$(x,u)(y,v) = (xy, uy + v + (x,y)\psi).$$

The obvious injection of U in E_ψ and projection of E_ψ onto Q are the non-trivial homomorphisms in the short exact sequence. The map sending $[\psi]$ to the equivalence class of the extension so constructed is well-defined and inverse to the map in the opposite direction, described above. So we have the very familiar one-to-one correspondence between $Z^2(Q,U)/B^2(Q,U)$ and the set of equivalence classes of extensions of U by Q with coupling χ. Accordingly, the second cohomology group $H^2(Q,U)$ of Q with coefficients in U may be viewed either as the former set under addition induced by pointwise composition of 2-cocycles, or as the latter set under *Baer addition* (see p.204 of [18]). We discuss this concept next.

Let

$$0 \to U \xrightarrow{\iota_1} G \xrightarrow{\pi_1} Q \to 1\ , \qquad 0 \to U \xrightarrow{\iota_2} H \xrightarrow{\pi_2} Q \to 1 \tag{2.1}$$

be extensions of U by Q, the equivalence classes of which correspond respectively to $[\psi_1] \in H^2(Q,U)$ and $[\psi_2] \in H^2(Q,U)$. We denote the subgroup $\{(u,-u) \mid u \in U\}$ of $U \times U$ by D and the subgroup $\{(g,h) \mid g \in G, h \in H, g\pi_1 = h\pi_2\}$ of $G \times H$ by E. Then

$$0 \to U \xrightarrow{\iota} E/D(\iota_1 \times \iota_2) \xrightarrow{\pi} Q \to 1 \tag{2.2}$$

is an extension of U by Q with coupling χ, where ι and π are defined by $u\iota = (u\iota_1, 1)D(\iota_1 \times \iota_2)$ and $((g,h)D(\iota_1 \times \iota_2))\pi = g\pi_1$ for $u \in U$ and $(g,h) \in E$. The extension (2.2) is called the *Baer sum* of the extensions (2.1). A 2-cocycle associated with (2.2) is cohomologous to $\psi_1 + \psi_2$. Furthermore, replacement of the extensions (2.1) by equivalent extensions in the formation of the Baer sum produces an extension equivalent to (2.2). So there is an obvious definition of Baer addition in $H^2(Q,U)$.

Most of the following ideas, leading up to Theorem 2.1, appear in §4, pp.66-71 of [16]. We say that $(\lambda, \tau) \in \text{Aut}(U) \times \text{Aut}(Q)$ is a *compatible pair* for χ if

$$((u)x\chi)\lambda = (u\lambda)x\tau\chi$$

for all $u \in U$ and $x \in Q$. The set of all compatible pairs for χ, denoted $\text{Comp}(\chi)$, is a subgroup of $\text{Aut}(U) \times \text{Aut}(Q)$. There is a right action of $\text{Comp}(\chi)$ on $Z^2(Q,U)$ defined by

$$(x,y)\psi^{(\lambda,\tau)} = (x\tau^{-1}, y\tau^{-1})\psi\lambda$$

for $x, y \in Q$ and $(\lambda, \tau) \in \mathrm{Comp}(\chi)$. This action leaves $B^2(Q,U)$ invariant, and so induces a right action of $\mathrm{Comp}(\chi)$ on $H^2(Q,U)$.

Now suppose that there is an isomorphism between the extensions (2.1) leaving U invariant; that is, there is an isomorphism $\theta: G \to H$ satisfying $U\iota_1\theta = U\iota_2$. (For example, an equivalence between the extensions is certainly an isomorphism of this kind.) Choose a normalised transversal function $\sigma_1: Q \to G$ for π_1 and let ψ_1 be the 2-cocycle defined relative to σ_1. Set

$$\lambda = \iota_1\theta\iota_2^{-1}, \qquad \tau = \sigma_1\theta\pi_2. \tag{2.3}$$

Then $(\lambda, \tau) \in \mathrm{Comp}(\chi)$, and $\sigma_2 = \tau^{-1}\sigma_1\theta$ is a normalised transversal function for π_2. If ψ_2 is the 2-cocycle defined relative to σ_2 then it is readily checked that $\psi_2 = \psi_1^{(\lambda,\tau)}$. Conversely, let (λ, τ) be an arbitrary element of $\mathrm{Comp}(\chi)$ and choose $[\psi] \in H^2(Q,U)$. Then the map defined by

$$(x, u) \mapsto (x\tau, u\lambda)$$

for $(x, u) \in E_\psi$, is an isomorphism of E_ψ onto $E_{\psi^{(\lambda,\tau)}}$ leaving U invariant. We have derived the following theorem (cf. (4.1)*, p.68 of [16]).

Theorem 2.1 *Let $[\psi_1], [\psi_2] \in H^2(Q,U)$. An extension in the equivalence class corresponding to $[\psi_1]$ is isomorphic, by an isomorphism leaving U invariant, to an extension in the equivalence class corresponding to $[\psi_2]$, if and only if $[\psi_1]$ and $[\psi_2]$ lie in the same orbit under the action of* $\mathrm{Comp}(\chi)$.

The U-invariance condition in Theorem 2.1 cannot be relaxed, in general. However, as we will demonstrate, Theorem 2.1 is sufficient for the purpose of answering the restricted isomorphism question.

For the rest of this chapter, T is a transitive 2-subgroup of S_4. In defining second cohomology of T with coefficients in a submodule of B, $t\chi$ is always conjugation by $t \in T$ in that submodule.

The proof of the next lemma involves calculating the cohomology of a finite cyclic group. This is done by means of the well-known description given in Proposition 7.1, p.201 of [12].

Lemma 2.2 *If $|T| = 4$ then $H^i(T, B) = 0$ for all $i \geq 1$. If $|T| = 8$ then $H^{2i}(T, B) = 0$ and $H^{2i-1}(T, B) \cong \mathbb{Z}_2$, for all $i \geq 1$.*

Proof. For a subgroup S of T, let $\mathrm{Ind}_S^T M$ denote the T-module induced from the S-module M (see p.67 of [4]). We have $B = \mathrm{Ind}_1^T C_{2^\infty}$ when $|T| = 4$, and $B = \mathrm{Ind}_{\langle ac\rangle}^T C_{2^\infty}$ when $|T| = 8$, treating C_{2^∞} as a trivial $\langle ac\rangle$-module. Calculating cohomology of the cyclic group $\langle ac\rangle$ with (trivial) coefficients in C_{2^∞}, we find that $H^{2i}(\langle ac\rangle, C_{2^\infty}) = 0$ and $H^{2i-1}(\langle ac\rangle, C_{2^\infty}) \cong \mathbb{Z}_2$, for $i \geq 1$. The conclusion then follows from Shapiro's Lemma (see Proposition 5.9, p.70 and Proposition 6.2, p.73 of [4]). □

Remark 2.3 Let δ be the derivation from D to B defined by $ac^i\delta = x_0$ and $c^i\delta = 0$, $1 \leq i \leq 4$, where $x_0 = (-1, -1, -1, -1)$ as always. It is easy to verify that there is no element z of B such that $t\delta = z^{1-t}$ for all $t \in D$. Thus, δ is not an inner derivation; that is, δ is a representative of the nonzero element of $H^1(D, B)$.

We use implicitly the well-known one-to-one correspondence between $H^1(Q, U)$ and the set of conjugacy classes of complements of U in $U \rtimes Q$ (see 11.1.3 and the preceding discussion on p.305 of [17]).

Proposition 2.4 *Let G and H be finite subgroups of BT such that $BG = BH = BT$ and $B \cap G = B \cap H \neq 1$. Set $Q = T$ and $U = B \cap G$ in* (2.1). *Further suppose that $\iota_1 = \iota_2$ and π_1, π_2 are both canonical projection onto T. With these assumptions, if the extensions* (2.1) *are equivalent then G and H are BT-conjugate.*

Proof. Set $B \cap G = B_1$. Note that ι_1 is a T-automorphism of B_1, by the very definition of the coupling χ.

The Baer sum of the equivalence classes of extensions corresponding to $[\psi_1]$ and $-[\psi_2]$ has representative

$$0 \to B_1 \xrightarrow{\iota} E/\overline{D} \xrightarrow{\pi} T \to 1,$$

where $\overline{D}$ is the normal subgroup $\{(z, z) \mid z \in B_1\}$ of E and ι, π are as defined after (2.2), with D replaced by $\overline{D}$. This extension splits, by hypothesis, and so we choose a complement $K/\overline{D}$ of $(B_1 \times B_1)/\overline{D} = B_1\iota$ in $E/\overline{D}$.

Let π_i also denote the extension of π_i to all of BT (of course $\pi_1 = \pi_2$ on BT). Define

$$\mathcal{E} = \{(e, f) \mid e \in BT, f \in BT, e\pi_1 = f\pi_2\} \quad \text{and} \quad \mathcal{D} = \{(z, z) \mid z \in B\}.$$

Note that $\mathcal{E}/\mathcal{D} \cong BT$; for instance, the homomorphism $\mu\colon BT \to \mathcal{E}/\mathcal{D}$ defined by $e\mu = (e, e\pi_1)\mathcal{D}$ is an isomorphism. Clearly $K\mathcal{D}/\mathcal{D}$ is a complement of $(B \times B)/\mathcal{D} \cong B$ in $\mathcal{E}/\mathcal{D}$. If $|T| = 4$ then there is a single conjugacy class of such complements, by Lemma 2.2, and this has representative $T\mu$. Thus $K^{(e,f)}\mathcal{D}/\mathcal{D} = T\mu$ for some fixed element (e, f) of $\mathcal{E}$. In other words, for each element $(g, h) \in K$, we have $g^e = h^f$. Since $K/\overline{D}$ complements $(B_1 \times B_1)/\overline{D}$ in $E/\overline{D}$ and B_1 is invariant under conjugation by elements of BT, it follows that $G^e = H^f$.

If $T = D$ then by Lemma 2.2 there are precisely two conjugacy classes of complements of $(B \times B)/\mathcal{D} \cong B$ in $\mathcal{E}/\mathcal{D}$. With reference to Remark 2.3, we see that one of these classes has representative $D\mu$ and the other has representative $\langle ax_0, c\rangle\mu$. If $K\mathcal{D}/\mathcal{D}$ lies in the former class then the same reasoning as above goes through. Otherwise, there is $(e, f) \in \mathcal{E}$ such that for each element (g, h) of K we have $g^e = h^f x_0 =$

$(hx_0)^f \in H^f$, since $x_0 \in B_1 \neq 1$ (see the comment after Proposition 1.1.3). Hence, we obtain the desired conclusion in this case also. □

The next theorem plays a prominent role in our solution of the extension problem, set out in Section 3.2.

Theorem 2.5 *Let $B_1 \neq 1$ be a finite T-submodule of B. Then $|H^2(T, B_1)|$ is the number of BT-conjugacy classes of finite 2-subgroups G of BS_4 such that $B \cap G = B_1$ and $BG = BT$.*

Proof. The number of BT-conjugacy classes of finite 2-subgroups G of BS_4 such that $B \cap G = B_1$ and $BG = BT$ is the same as the number of conjugacy classes of complements of B/B_1 in $BT/B_1 \cong (B/B_1) \rtimes T$, namely $|H^1(T, B/B_1)|$. Now the short exact sequence

$$0 \to B_1 \to B \to B/B_1 \to 0$$

gives rise to a long exact sequence

$$\cdots \to H^1(T,B) \xrightarrow{\gamma} H^1(T,B/B_1) \to H^2(T,B_1) \to H^2(T,B) \to \cdots,$$

where γ is induced by the natural surjection of B onto B/B_1 (see (2.3), p.189 of [12]). If $|T| = 4$ then the result follows from Lemma 2.2. If $T = D$ then by Lemma 2.2 and the long exact sequence, $H^1(T, B/B_1)/\mathrm{im}\gamma \cong H^2(T, B_1)$. But since $x_0 \in B_1$, γ maps the 1-cohomology class of δ (as defined in Remark 2.3) to zero. Thus $H^1(D, B/B_1) \cong H^2(D, B_1)$ and we are done. □

Proposition 2.4 and Theorem 2.5 yield the following.

Proposition 2.6 *Let $\{G_1, \ldots, G_n\}$ be a complete and irredundant set of BT-conjugacy class representatives of the finite 2-subgroups G of BS_4 such that $BG = BT$ and $B \cap G = B_1$ for some finite T-submodule B_1 of B. Then*

$$\{0 \to B_1 \overset{\text{inc.}}{\to} G_i \overset{\text{proj.}}{\to} T \to 1 \mid 1 \leq i \leq n\}$$

is a complete and irredundant set of representatives of the elements of $H^2(T, B_1)$.

Informally, BT-conjugacy is the same concept as equivalence of extensions in this setting. As noted earlier, equivalence is an isomorphism between extensions that maps diagonal subgroup onto diagonal subgroup. In Chapter 6, we will see that $GL(4)$-conjugacy captures all such isomorphisms.

The last result in this chapter is required for applications to the restricted isomorphism question and conjugacy problem.

Theorem 2.7 *Suppose that G, H are finite subgroups of BS_4 such that $BG = BH = BT$. If θ is an isomorphism of G onto H with $(B \cap G)\theta = B \cap H$, then $(B \cap G)^s = B \cap H$ for some $s \in \mathsf{N}_{S_4}(T)$. In particular, if $|T| = 8$ then $B \cap G = B \cap H$.*

Proof. First suppose that $|T| = 4$. There is $\tau \in \mathrm{Aut}(T)$ such that

$$z^t\lambda = z\lambda^{t\tau}$$

for all $z \in B \cap G$ and $t \in T$, where λ is θ restricted to $B \cap G$ (cf. the reasoning framing (2.3)). Since τ must act as conjugation by some $s \in \mathsf{N}_{S_4}(T)$, this implies that the composite of λ with conjugation by s^{-1} is a T-module isomorphism of $B \cap G$ onto $(B \cap H)^{s^{-1}}$. Hence $B \cap G = (B \cap H)^{s^{-1}}$ by (the remark after) Proposition 1.1.2, verifying the claim for $|T| = 4$.

Suppose now that $T = D$. Set $G_1 = BC \cap G$ and $G_2 = G_1\theta$. Then $B \cap G_1 = B \cap G$, $BG_1 = BC$ and $B \cap G_2 = B \cap H$. Since $G_2/B \cap G_2 \cong C$ and C is the unique cyclic subgroup of order 4 in D, it follows that $BG_2 = BC$. By the first paragraph, there is $s \in \mathsf{N}_{S_4}(C) = D$ such that $(B \cap G_1)^s = B \cap G_2$. Thus $B \cap G = B \cap H$. □

Chapter 3

The case $T = V_4$

3.1 The finite V_4-submodules of B

Let $A = (\mathbb{Z}/2^n\mathbb{Z})V_4$, where n is fixed but arbitrarily large. Our first objective in this section is to develop an understanding of $\mathcal{H}(A_A)$. After the usual translation (cf. Section 1.2), this will then allow us to list the finite V_4-submodules of B, and some of the inclusion relations between them (the full Hasse diagram is not needed for this purpose).

The analysis rests upon distinguishing between the *cyclic* submodules of A_A (those which are cyclic as additive groups) and the noncyclic submodules. It will be seen that the noncyclic submodules correspond to the submodules of a direct sum, and so the theory of [10] can again be brought to bear.

We begin by determining the noncyclic submodules of smallest order. For each involution $t \in V_4$, let F_t be the submodule of A_A generated by $2^{n-1}(1+t)$. Note that $\mathrm{C}_{V_4}(F_t) = \langle t \rangle$.

Lemma 3.1.1 *The noncyclic submodules of A_A of order 4 are the F_t.*

Proof. The set of isomorphism classes of noncyclic A-modules of order 4 is in one-to-one correspondence with the set of equivalence classes of representations of V_4 in $\mathrm{Aut}(C_2 \times C_2) \cong S_3$. The latter set has three elements, implying by Proposition 1.1.2 that there are at most three distinct noncyclic submodules of A_A of order 4. By construction, there are precisely three such submodules, and they are the F_t. □

For each involution $t \in V_4$, define

$$F_t^+ = \langle 1+t, s+st \rangle / F_t \qquad \text{and} \qquad F_t^- = \langle 1-t, s-st \rangle / F_t,$$

where $s \neq t$ is an involution of V_4. The V_4-modules F_t^+ and F_t^- are respectively fixed and inverted elementwise by t.

Proposition 1.1.4 is used without reference in proofs of the next few results.

Lemma 3.1.2 *In the notation above,*

$$F_t^{\perp}/F_t = F_t^+ \oplus F_t^-.$$

Furthermore, $\mathcal{L}(F_t^+)$ *and* $\mathcal{L}(F_t{}^-)$ *are isomorphic to the submodule lattice of the regular* $(\mathbb{Z}/2^{n-1}\mathbb{Z})C_2$*-module, and any two of the* $\mathcal{L}(F_t^{\perp}/F_t)$ *have isomorphic Hasse diagrams.*

Proof. Certainly $F_t^{\perp} = \mathrm{ann}_A F_t$ contains $1+t$ and $1-t$. The submodule of A_A generated by these elements has order $2^{4n-2} = |A|/|F_t| = |F_t^{\perp}|$, and so is $F_t^{\perp}$. Then the stated direct decomposition is clear. Since t acts as a scalar on F_t^+ and F_t^-, which as $(\mathbb{Z}/2^{n-1}\mathbb{Z})\langle s\rangle$-modules are isomorphic to the regular module, we obtain the first part of the second statement. Now let t and t' be distinct involutions of V_4. We get an abelian group isomorphism between $F_t^{\perp}$ and $F_{t'}^{\perp}$ by linearly extending the automorphism $t \leftrightarrow t'$ of V_4. This induces an inclusion-preserving bijection between the set of V_4-submodules of $F_t^{\perp}$ containing F_t, and the set of V_4-submodules of $F_{t'}^{\perp}$ containing $F_{t'}$. So we have proved the last statement in its entirety. □

Proposition 3.1.3 *Let* M *be a submodule of* A_A*. Then* M *is neither cyclic nor the dual of a cyclic submodule if and only if* $F_t \leq M \leq F_t^{\perp}$ *for some involution* $t \in V_4$*.*

Proof. One direction is obvious. Conversely, suppose that M is neither cyclic nor the dual of a cyclic submodule. Then $M \geq F_t$ for some t and also $M^{\perp} \geq F_s$ for some s. We will show that it is possible to choose $t = s$.

First, $\Omega_1(A) \cong \mathbb{F}_2 V_4$ has a unique maximal V_4-submodule, by Proposition 1.1.1. Thus A_A has a unique elementary abelian submodule E of order 8, namely the sum of any two of the F_t, and so we may assume henceforth that $M \not\geq E$ and $M \not\leq E^{\perp}$. This means that M and $M^{\perp}$ (as abelian groups) both have rank 2, and consequently $M \cong C_{2^n} \times C_{2^n}$.

Multiplication by 2^{n-1} is an endomorphism of each submodule of A_A, so that $A_A/2A_A \cong \Omega_1(A)$ and $(M+2A_A)/2A_A \cong 2^{n-1}M = F_t$. For any submodule U of order 4 in $\Omega_1(A)$, we have $U \cong \Omega_1(A)/U$. Therefore,

$$F_t \cong \Omega_1(A)/F_t \cong A_A/(M+2A_A).$$

This demonstrates that $(M+2A_A)^{\perp}$ has order 4, hence is singly generated, and so $F_t \cong A_A/(M+2A_A) \cong (M+2A_A)^{\perp}$. By Proposition 1.1.2, $F_t = (M+2A_A)^{\perp}$. Therefore $F_t \leq M^{\perp}$ and we are done. □

Proposition 3.1.3 enables us to divide the listing of submodules of A_A into two separate parts. The first of these is listing, for each t, the submodules that contain F_t and are contained in $F_t^{\perp}$: the elements of $\mathcal{V}(F_t^{\perp}/F_t)$, in other words. The second part is listing the cyclic submodules and their duals. Likewise, listing the elements of the edge set $\mathcal{E}(A_A)$ of $\mathcal{H}(A_A)$ is a bipartite procedure. To see this, first note that the submodules of A_A that are either cyclic or contain a cyclic maximal submodule form a meet-semilattice. Denote this meet-semilattice by $\mathcal{C}$. By taking the dual of each element of $\mathcal{C}$, we obtain a join-semilattice $\mathcal{C}^\perp$. Of course, the Hasse diagrams of $\mathcal{C}$ and $\mathcal{C}^\perp$ are dual.

Proposition 3.1.4 *If $W_1/W_2 \in \mathcal{E}(A_A)$ then W_1/W_2 is an edge in the Hasse diagram of $\mathcal{C}$, $\mathcal{C}^\perp$, or $\mathcal{L}(F_t^\perp/F_t)$ for some involution $t \in V_4$.*

Proof. If W_2 is cyclic (respectively, dual of a cyclic submodule) then $W_1 \in \mathcal{C}$ (respectively, $W_1 \in \mathcal{C}^\perp$); this statement also holds with "W_1" and "W_2" interchanged. Suppose, then, that neither W_1 nor W_2 is cyclic or the dual of a cyclic submodule. By Proposition 3.1.3, $F_t \leq W_1 \leq F_t^\perp$ and $F_s \leq W_2 \leq F_s^\perp$ for some t and s. If $W_2 \geq F_t$ then we are done. Otherwise, $W_1 = W_2 + F_t \leq F_s^\perp$, completing the proof. □

By Lemma 3.1.2 and Proposition 3.1.4, to describe $\mathcal{H}(A_A)$ we only need to describe the Hasse diagram of $\mathcal{C}$, and $\mathcal{H}(F_t^\perp/F_t)$ for a single t. We have already indicated the procedure to be followed in describing the latter Hasse diagram. In the listing of submodules, the former Hasse diagram may be ignored. This is by way of the next result, which justifies concentrating on noncyclic submodules in this section. (Actually, some detail of the Hasse diagram of $\mathcal{C}$ is needed later on—see the proof of Theorem 3.3.10—and it is more convenient to develop the required results there.)

Proposition 3.1.5 *Let G be a finite irreducible 2-subgroup of BS_4 such that $BG = BV_4$ and $B \cap G$ is cyclic. Then there is a $GL(4)$-conjugate of G in BV_4 with noncyclic diagonal subgroup.*

Proof. By the sentence immediately preceding Theorem 2 of [3], G is isomorphic to one of two known groups, each of which has a noncyclic maximal abelian normal subgroup with quotient V_4. The result then follows from Proposition 1.3.6. □

In summary: if $\mathcal{L}(F_t^\perp/F_t)$ for $t = a$ (say) is known, then the poset of finite noncyclic V_4-submodules of B is known. An edge in the Hasse diagram of this poset corresponds to an edge in at least one of the $\mathcal{H}(F_t^\perp/F_t)$, by Proposition 3.1.4. We can pass to the atlas of $\mathcal{L}(F_t^\perp/F_t)$ from that of $\mathcal{L}(F_a^\perp/F_a)$ by applying the lattice isomorphism which arises from conjugation by the appropriate element of $\mathrm{Aut}(V_4) < S_4$, as in the proof of Lemma 3.1.2. This reduction points up an important feature of our classification of the finite irreducible 2-groups in BV_4. Conjugacy by an element of S_4 is an equivalence relation on the set of those groups, and the action leaves diagonal subgroups invariant. Eventually, we will produce a complete list of representatives of the resultant equivalence classes, with the property that distinct diagonal subgroups of representatives are pairwise non-conjugate in BS_4. The production of this list is begun in Proposition 3.1.12 and completed in Theorem 3.3.1.

We will specify a finite V_4-submodule of B by a group generating set. The diagonal matrices which appear in these generating sets are defined for $k \geq 0$ by

$$x_k = (\omega_k, \omega_k, \omega_k, \omega_k), \quad y_k = (\omega_k, \omega_k, \omega_k^{-1}, \omega_k^{-1}), \quad u_k = y_k^{(2,3)}, \; v_k = y_k^{(1,3)},$$

where $\omega_k = \exp(2^{-k}\pi\sqrt{-1})$ as in Section 1.2. Note that $x_0 = y_0 = u_0 = v_0 = x_1y_1u_1v_1$. Denote by X the V_4-submodule $\langle x_0, x_1, \ldots \rangle$ of B (the scalars of $GL(4)$ in B) and by Y the V_4-submodule $\langle y_0, y_1, \ldots \rangle$ of B. Set $U = Y^{(2,3)}$ and $V = Y^{(1,3)}$. Recall that $\langle (1,2), (2,3) \rangle$ is a complement of V_4 in S_4 and acts as the automorphism group of V_4 by conjugation; it also acts as S_3 on $\{Y, U, V\}$. We see that $C_{V_4}(Y) = \langle a \rangle$ and $t \in V_4 \backslash \langle a \rangle$ inverts Y elementwise; the corresponding statements obtained after replacing Y by U or V and a by b or ab, respectively, are also clear.

The next lemma collects together some (easily proved) facts about the V_4-modules just defined. Recall that $B_0 = \langle x_0 \rangle$.

Lemma 3.1.6 *Set $C_1 = X$, $C_a = Y$, $C_b = U$ and $C_{ab} = V$. Then $B = \prod_{t \in V_4} C_t$ and $B \cap SL(4) = \prod_{t \in V_4 \backslash \{1\}} C_t$. Also, for distinct $t_i \in V_4$,*

(i) $C_{t_1} \cap C_{t_2} = \Omega_1(C_{t_1}) = \Omega_1(C_{t_2}) = B_0$,

(ii) $C_{t_1}C_{t_2} \cap C_{t_3} = \Omega_1(C_{t_3}) = B_0$,

(iii) $C_{t_1}C_{t_2} \cap C_{t_3}C_{t_4} = \Omega_1(C_{t_1}C_{t_2}) = \Omega_1(C_{t_3}C_{t_4})$,

(iv) $C_{t_1}C_{t_2}C_{t_3} \cap C_{t_4} = \Omega_2(C_{t_4})$.

A finite irreducible subgroup G of $GL(4)$ has centre $Z(G)$ consisting solely of scalar matrices; if $G \leq BS_4$ then $Z(G) = G \cap X$.

Translating Lemma 3.1.1, we see that the noncyclic V_4-submodules of B of order 4 are

$$F_a = \langle x_1y_1, u_1v_1 \rangle, \qquad F_b = \langle x_1u_1, y_1v_1 \rangle, \qquad F_{ab} = \langle x_1v_1, y_1u_1 \rangle,$$

using the same notation as that used earlier for the matching submodules of A_A. For the reasons mentioned after Propositions 3.1.4 and 3.1.5, in the rest of this section we restrict attention to the finite V_4-submodules of B containing F_a.

From Lemma 3.1.6, we deduce that

$$B/F_a = (XY/F_a) \times (UV/F_a). \tag{3.1}$$

This direct decomposition of the V_4-module B/F_a is an analogue of the direct decomposition of $F_a^{\perp}/F_a$ given by Lemma 3.1.2, with XY/F_a and UV/F_a matching

F_a^+ and F_a^-, respectively. The methods of [10] may be applied to (3.1) once the V_4-module structure of the factors, and the Hasse diagrams of their submodule lattices, are known. This information is available after translation of earlier results in this chapter and Chapter 1, and is given below.

By Lemma 3.1.2, Proposition 1.2.2 and the subsequent discussion in Section 1.2, the finite V_4-submodules of XY properly containing F_a are defined by

$$F_a^+(i,j,\delta_1) = \langle x_i, y_j, (x_{i+1}y_{j+1})^{\delta_1}\rangle,$$

where i,j range over the positive integers and $\delta_1 \in \{0,1\}$. Note that F_a itself may be included by allowing $i = j = 0$ when $\delta_1 = 1$. Similarly, the finite V_4-submodules of UV properly containing F_a are the $F_a^-(k,l,\delta_2)$ for $k,l \geq 1$ and $\delta_2 \in \{0,1\}$, where

$$F_a^-(k,l,\delta_2) = \langle u_k, v_l, (u_{k+1}v_{l+1})^{\delta_2}\rangle;$$

F_a is included exceptionally as $F_a^-(0,0,1)$.

The edge sets $\mathcal{E}(XY/F_a)$ and $\mathcal{E}(UV/F_a)$ may be written down with reference to the discussion at the end of Section 1.2 (see Figures 1.1 and 1.2 in particular). The set $\mathcal{E}(XY/F_a)$ consists of the edges $F_a^+(i+1,j,0)/F_a^+(i,j,0)$, $F_a^+(i,j,1)/F_a^+(i,j,0)$, $F_a^+(i,j+1,0)/F_a^+(i,j,0)$ and $F_a^+(i+1,j+1,0)/F_a^+(i,j,1)$, with i,j ranging over the non-negative integers as indicated above. Then $\mathcal{E}(UV/F_a)$ is obtained by replacing the superscript $+$ with $-$ in the description just given.

Each submodule in our list is defined by a unique *label*: a string of integer parameters specifying a group generating set for the submodule. The following convenient notation is used in stating ranges on those parameters.

Notation 3.1.7 For integers $i_1,\ldots,i_m$, we write $i_1,\ldots,i_m \gtrsim 0$ to mean that either $i_1,\ldots,i_m \geq 1$ or $i_1 = \cdots = i_m = 0$.

Following [10], we call a section of a finite submodule of a module *low*. By Theorem 3.1 of [10] and (3.1), the finite V_4-submodules of B containing F_a are in one-to-one correspondence with the isomorphisms from low sections of XY/F_a onto low sections of UV/F_a. Using this correspondence, we now begin construction of the submodules (cf. the proof of Proposition 1.2.2).

The submodule of B corresponding to the isomorphism

$$F_a^+(i,j,\delta_1)/F_a^+(i,j,\delta_1) \to F_a^-(k,l,\delta_2)/F_a^-(k,l,\delta_2)$$

between zero low sections of XY/F_a and UV/F_a will be denoted $F(i,j,k,l,\delta_1,\delta_2)$; in terms of group generators,

$$F(i,j,k,l,\delta_1,\delta_2) = \langle x_i, y_j, u_k, v_l, (x_{i+1}y_{j+1})^{\delta_1}, (u_{k+1}v_{l+1})^{\delta_2}\rangle. \tag{3.2}$$

We regard these submodules as forming four *families*. In labels $F(i,j,k,l,\delta_1,\delta_2)$, the parameters δ_1,δ_2 range independently over $\{0,1\}$, and together they identify the family concerned. The parameters i,j,k,l, determine the order of a submodule in a given family, and range according to the values of δ_1 and δ_2. Specifically, $i,j \geq 1$ when $\delta_1 = 0$ and $i,j \gtrsim 0$ when $\delta_1 = 1$; similarly, $k,l \geq 1$ when $\delta_2 = 0$ and $k,l \gtrsim 0$ when $\delta_2 = 1$. To avoid overlong expressions, we sometimes write δ for (the "vector" with "coordinates") δ_1,δ_2, and $F(i,j,k,l,\delta)$ will be used informally as a generic label for members of the four families. With respect to (3.1), the submodules with these labels are the finite Cartesian submodules of B containing F_a.

Next, we consider the (non-Cartesian) submodules corresponding to isomorphisms between simple low sections of XY/F_a and UV/F_a. Of course, there is a unique isomorphism from each of these sections to any other. For given i and j, there are three simple sections of the form $W/F_a^+(i,j,0)$ in XY/F_a, namely those with W one of $F_a^+(i,j,1)$, $F_a^+(i,j+1,0)$, or $F_a^+(i+1,j,0)$. We will use i, j, and a parameter ξ_1 ranging over $\{-1,0,1\}$ (in the same order as the listing of possibilities for W) to index these. Similarly, given k and l, there are three simple sections of the form $W/F_a^-(k,l,0)$ in UV/F_a, namely those with W one of $F_a^-(k,l,1)$, $F_a^-(k,l+1,0)$, or $F_a^-(k+1,l,0)$. We will use k, l, and a parameter ξ_2 ranging over $\{-1,0,1\}$ to index these. The submodule corresponding to the isomorphism from a simple section indexed i,j,ξ_1 onto a section indexed k,l,ξ_2 will be called $F(i,j,k,l,0,0,\xi_1,\xi_2)$; in terms of group generators,

$$F(i,j,k,l,0,0,\xi_1,\xi_2) = \langle x_i, y_j, u_k, v_l, x_{i+1}^{\xi_1} y_{j+1}^{\xi_1'} u_{k+1}^{\xi_2} v_{l+1}^{\xi_2'} \rangle, \tag{3.3}$$

where ξ_1' and ξ_2' are representatives in $\{-1,0,1\}$ of the congruence class modulo 3 of $1-\xi_1$ and $1-\xi_2$, respectively. With ξ_1 and ξ_2 ranging independently over $\{-1,0,1\}$, these submodules form nine more families. Given that here $\delta_1 = \delta_2 = 0$, the parameters i,j,k,l range (independently of ξ_1 and ξ_2) over the positive integers. For brevity, ξ is sometimes written in place of ξ_1,ξ_2, and $F(i,j,k,l,0,0,\xi)$ is an informal generic label for members of the nine families.

There is only one other kind of simple low section in XY/F_a, and it is of the form $F_a^+(i+1,j+1,0)/F_a^+(i,j,1)$; its analogue in UV/F_a is $F_a^-(k+1,l+1,0)/F_a^-(k,l,1)$. The submodule corresponding to the isomorphism between the first of these sections and the simple section indexed by k,l,ξ_2 above will be called $F(i,j,k,l,1,0,\xi_2)$. The submodule corresponding to the isomorphism between the second of these sections and the simple section indexed by i,j,ξ_1 above will be called $F(i,j,k,l,0,1,\xi_1)$. The submodule corresponding to the isomorphism of $F_a^+(i+1,j+1,0)/F_a^+(i,j,1)$ onto $F_a^-(k+1,l+1,0)/F_a^-(k,l,1)$ will be called $F(i,j,k,l,1,1,1)$. In terms of group generators,

$$F(i,j,k,l,1,0,\xi) = \langle x_{i+1}y_{j+1}, y_j, u_k, v_l, x_{i+1}u_{k+1}^{\xi} v_{l+1}^{\xi'} \rangle, \tag{3.4}$$

$$F(i,j,k,l,0,1,\xi) = \langle x_i, y_j, u_{k+1}v_{l+1}, v_l, x_{i+1}^{\xi} y_{j+1}^{\xi'} u_{k+1} \rangle, \tag{3.5}$$
$$F(i,j,k,l,1,1,1) = \langle x_{i+1}y_{j+1}, y_j, u_{k+1}v_{l+1}, v_l, x_{i+1}u_{k+1} \rangle, \tag{3.6}$$

where ξ' is the representative in $\{-1,0,1\}$ of the congruence class mod 3 of $1-\xi$. These submodules form seven more families, and in their labels the ranges of i,j,k,l depend only on δ.

We will use $F(i,j,k,l,\delta,\xi)$ as a generic label for members of the twenty families of submodules constructed so far. In the four Cartesian families, ξ stands for a void. In the first nine non-Cartesian families, δ is $0,0$ and ξ stands for ξ_1, ξ_2 as explained above. In the next six non-Cartesian families, δ is either $1,0$ or $0,1$ and ξ stands for a single parameter ranging over $\{-1,0,1\}$. In the last family, δ, ξ is $1,1,1$. (These conventions are in line with the use of "A.B.C." as a generic name for a person without excluding people who have no middle name or more than one middle name.)

Finally, it remains to deal with the submodules corresponding to isomorphisms between nonsimple low sections. Since a acts trivially on XY/F_a but invertingly on UV/F_a, isomorphic sections of XY/F_a and UV/F_a must be elementary abelian. In a module which as abelian 2-group has rank 2, a nonsimple low elementary abelian section is the quotient of a finite submodule over its multiple by the scalar 2. In XY/F_a, the relevant sections are the $F_a^+(i+1,j+1,\delta_1)/F_a^+(i,j,\delta_1)$, and in UV/F_a, they are the $F_a^-(k+1,l+1,\delta_2)/F_a^-(k,l,\delta_2)$.

When δ_1 or δ_2 is 0, the section is V_4-trivial. Consequently, for fixed i,j,k,l there are $|\mathrm{Aut}(C_2 \times C_2)| = 6$ isomorphisms

$$F_a^+(i+1,j+1,0)/F_a^+(i,j,0) \to F_a^-(k+1,l+1,0)/F_a^-(k,l,0).$$

Instead of representing such an isomorphism by a 2×2 matrix over $\mathbb{F}_2$ (entailing the use of four more parameters), we use two parameters, α_1 and α_2, ranging over $\{-1,0,1\}$ subject to $\alpha_1 \neq \alpha_2$. In effect, one of these represents the three choices for first row of the matrix, the other represents the two choices for its second row. Then the corresponding submodules of B are

$$F(i,j,k,l,0,0,2,\alpha_1,\alpha_2) = \langle x_i, y_j, u_k, v_l, x_{i+1}u_{k+1}^{\alpha_1}v_{l+1}^{\alpha_1'}, y_{j+1}u_{k+1}^{\alpha_2}v_{l+1}^{\alpha_2'} \rangle, \tag{3.7}$$

where α_1' and α_2' are representatives in $\{-1,0,1\}$ of the congruence class modulo 3 of $1-\alpha_1$ and $1-\alpha_2$, respectively. We regard these submodules as forming six families, according to the values of α_1 and α_2. The informal generic label for submodules is expanded to $F(i,j,k,l,\delta,\xi,\alpha)$. In the case of the twenty earlier families, α stands for a void, while in the case of the six new families, ξ stands for 2 (the composition length of domains of the corresponding isomorphisms) and α stands for the pair α_1, α_2.

The nonsimple elementary abelian sections with $\delta_1 = 1$ or $\delta_2 = 1$ are not trivial but uniserial V_4-modules; they are not isomorphic to any section considered before but they are all isomorphic to each other. For given i, j, k, l, there are precisely two V_4-isomorphisms

$$F_a^+(i+1, j+1, 1)/F_a^+(i, j, 1) \to F_a^-(k+1, l+1, 1)/F_a^-(k, l, 1),$$

and the corresponding submodules of B are

$$F(i, j, k, l, 1, 1, 2, \alpha) = \langle x_i, y_j, u_k, v_l, x_{i+1}y_{j+1}, x_{i+1}u_{k+1}, x_{i+2}^{\alpha}y_{j+2}u_{k+2}v_{l+2}\rangle, \tag{3.8}$$

where $\alpha \in \{-1, 1\}$. These submodules form two more families; in the informal generic label, ξ stands once more for 2. The construction process is now complete, and has given us 28 families in all.

Theorem 3.1.8 *The submodules defined in* (3.2)–(3.8), *with the parameters ranging as indicated, constitute a complete and irredundant list of the finite* V_4*-submodules of* B *containing* F_a.

For ease of reference, we provide now a summary of the labelling scheme. Each submodule in the list has a label beginning "$F(i, j, k, l, \delta_1, \delta_2$", where δ_1 and δ_2 range independently over $\{0, 1\}$ while i, j, k, l range over the non-negative integers, satisfying

$i, j \geq 1$ or $i, j \gtrsim 0$ according to whether δ_1 is 0 or 1 respectively, and

$k, l \geq 1$ or $k, l \gtrsim 0$ according to whether δ_2 is 0 or 1 respectively.

The label may then end with ")", in which case the submodule is Cartesian. Otherwise, the submodule is non-Cartesian and the label continues subject to conditions which depend on the values of δ_1, δ_2. Namely,

when this parameter pair is $0, 0$, the label may continue either with "$, \xi_1, \xi_2)$" where $\xi_1, \xi_2 \in \{-1, 0, 1\}$, or with "$, 2, \alpha_1, \alpha_2)$" where $\alpha_1, \alpha_2 \in \{-1, 0, 1\}$ and $\alpha_1 \neq \alpha_2$;

when it is $0, 1$ or $1, 0$, the label may continue with "$, \xi)$" where $\xi \in \{-1, 0, 1\}$;

when it is $1, 1$, the label may continue either with "$, 1)$" or with "$, 2, \alpha)$" where $\alpha \in \{-1, 1\}$.

In terms of group generators, the submodules are given by (3.2)–(3.8); for a parameter λ ranging over $\{-1, 0, 1\}$, λ' is the representative in that set of the congruence class modulo 3 of $1 - \lambda$.

It is worthwhile noting how some properties of the listed submodules can be read off their labels. First,

$$\begin{aligned} F(i, j, k, l, \delta, \xi, \alpha) \cap X &= \langle x_i \rangle, \\ F(i, j, k, l, \delta, \xi, \alpha) \cap Y &= \langle y_j \rangle, \\ F(i, j, k, l, \delta, \xi, \alpha) \cap U &= \langle u_k \rangle, \\ F(i, j, k, l, \delta, \xi, \alpha) \cap V &= \langle v_l \rangle. \end{aligned}$$

The *interval of definition* of a listed submodule is the minimal interval of $\mathcal{L}(B/F_a)$ with Cartesian endpoints that contains the submodule (see Section 3 of [10]). Specifically, if the submodule corresponds to an isomorphism $M_1/M_2 \to N_1/N_2$, then $M_2 \oplus N_2$ and $M_1 \oplus N_1$ are respectively the bottom and top endpoints of its interval of definition. Writing the submodule as $F(i,j,k,l,\delta,\xi,\alpha)$, we see that its interval of definition has bottom endpoint $F(i,j,k,l,\delta)$, which is the (unique) maximal Cartesian submodule of $F(i,j,k,l,\delta,\xi,\alpha)$. The top endpoint is determined by the bottom endpoint and ξ; its label may be written down after consulting the exposition leading up to Theorem 3.1.8. For example, $F(i,j,k,l,0,0,-1,-1)$ corresponds to the isomorphism

$$F_a^+(i,j,1)/F_a^+(i,j,0) \to F_a^-(k,l,1)/F_a^-(k,l,0),$$

and so its interval of definition has top endpoint $F(i,j,k,l,1,1)$. Of course, given the interval of definition of $F(i,j,k,l,\delta,\xi,\alpha)$—that is, given the values of i,j,k,l and ξ—we know the submodule's order: this is 2^σ if ξ is a void, $2^{\sigma+2}$ if $\xi = 2$, and $2^{\sigma+1}$ for all other values of ξ, where $\sigma = i+j+k+l+\delta_1+\delta_2$.

The following lemma enables us to recognise irreducibility of groups constructed in solving the extension problem.

Lemma 3.1.9 *The non-faithful V_4-modules in the list of Theorem* 3.1.8 *are precisely the $F(i,j,0,0,0,1)$ and $F(i,j,0,0,1,1)$.*

Proof. If M is in the list then $\mathsf{C}_{V_4}(M) \le \mathsf{C}_{V_4}(F_a) = \langle a \rangle$. Thus, if M is not faithful then $M \le XY$. The finite V_4-submodules of XY containing F_a were earlier called $F_a^+(i,j,\delta_1)$, and now appear with labels $F(i,j,0,0,\delta_1,1)$. □

Recall that the poset of finite noncyclic V_4-submodules of B is essentially known once $\mathcal{H}(B/F_a)$ is known. Having described $\mathcal{V}(B/F_a)$, then, we turn now to the problem of describing $\mathcal{E}(B/F_a)$. Section 5 of [10] provides the basis of our approach, as it did in Section 1.2.

We seek to construct an irredundant atlas of $\mathcal{L}(B/F_a)$, in the sense that the Hasse diagram of each page has at least one edge not in the Hasse diagram of any other page. Of course, the atlas should still be complete, and to this end will be comprised of some or all of the following: the sublattice of finite Cartesian submodules, the intervals of definition (to cover the restriction edges in $\mathcal{H}(B/F_a)$), and each page that arises from a pair of perspective edges in $\mathcal{H}(XY/F_a)$ or $\mathcal{H}(UV/F_a)$ (to cover the composition edges in $\mathcal{H}(B/F_a)$). The irredundancy requirement forces us to select for inclusion in the atlas only those intervals of definition that are *maximal* (each interval of definition is contained in at least one maximal, and no maximal is contained in any other). Observe that each simple low section of XY/F_a except $F_a^+(1,1,0)/F_a$ is a section of some $F_a^+(i+1,j+1,0)/F_a^+(i,j,0)$, with a similar statement holding for simple low sections of UV/F_a. Using this observation and knowledge of the submodule construction process, it may be deduced that the maximal intervals of definition in this case are of the form:

$$\begin{gathered}
\mathcal{L}(F(i+1,j+1,k+1,l+1,0,0)/F(i,j,k,l,0,0)),\\
\mathcal{L}(F(i+1,j+1,k+1,l+1,1,1)/F(i,j,k,l,1,1)),\\
\mathcal{L}(F(1,1,k+1,l,0,0)/F(0,0,k,l,1,0)),\\
\mathcal{L}(F(1,1,k,l+1,0,0)/F(0,0,k,l,1,0)),\\
\mathcal{L}(F(i+1,j,1,1,0,0)/F(i,j,0,0,0,1)),\\
\mathcal{L}(F(i,j+1,1,1,0,0)/F(i,j,0,0,0,1)).
\end{gathered}$$

Since $F(i+1,j+1,k+1,l+1,0,0)/F(i,j,k,l,0,0)$ is

$$(F_a^+(i+1,j+1,0)\oplus F_a^-(k+1,l+1,0))/(F_a^+(i,j,0)\oplus F_a^-(k,l,0)),$$

which is isomorphic (as a trivial V_4-module) to $V(2,2)\oplus V(2,2)$, we see that an interval of definition of the first listed form is isomorphic to $\mathcal{L}(4,2)$. An interval of the second listed form is isomorphic to $\mathcal{L}(C_4\times C_4)$, the subgroup lattice of $C_4\times C_4$. All other intervals are isomorphic to $\mathcal{L}(2,2)$.

Proposition 3.1.10 *If M and N are finite V_4-submodules of B containing F_a such that N is maximal in M, then M/N occurs as an edge in the Hasse diagram of at least one of the following:*

$$\mathcal{L}(F(i+1,j+1,k+1,l+1,0,0)/F(i,j,k,l,0,0))\cong\mathcal{L}(4,2), \tag{3.9}$$
$$\mathcal{L}(F(i+1,j+1,k+1,l+1,1,1)/F(i,j,k,l,1,1))\cong\mathcal{L}(C_4\times C_4), \tag{3.10}$$
$$\mathcal{L}(F(i+1,j+1,1,1,0,0)/F(i,j,0,0,0,1))\cong\mathcal{L}(3,2), \tag{3.11}$$
$$\mathcal{L}(F(1,1,k+1,l+1,0,0)/F(0,0,k,l,1,0))\cong\mathcal{L}(3,2), \tag{3.12}$$

for some i,j,k,l satisfying the usual constraints. Furthermore, the atlas of $\mathcal{L}(B/F_a)$ comprised of all pages of the forms (3.9)–(3.12) *is irredundant.*

Proof. Our atlas is constructed according to the discussion above. General descriptions of the two types of perspectivity pages are given in (v) and (vi) of Theorem 4.3 in [10], and we will use these to determine specifically the perspectivity pages in this atlas (such pages are designated "type (v)" or "type (vi)" accordingly). Figures 1.1 and 1.2 should be referred to in the rest of the proof.

Pages of the forms (3.9) and (3.10) are maximal intervals of definition. It is easily seen that any other maximal interval of definition is contained in a page (3.11) or a page (3.12). These lattices are isomorphic to $\mathcal{L}(3,2)$ by reasoning similar to that which established the isomorphism in (3.9). We claim that pages of the forms (3.11) and (3.12) are the irredundant perspectivity pages.

Denote by $\mathcal{S}$ the set consisting of low sections of XY/F_a that are isomorphic to sections of UV/F_a. To determine the pages of type (v), we need to list all $S_1/S_2\in\mathcal{S}$

with the property that S_1 has a maximal submodule not containing S_2. An element of $\mathcal{S}$ has composition length at most 2. If S_1/S_2 has length 2 then certainly each maximal submodule of S_1 contains S_2. The simple elements of $\mathcal{S}$ with the required property are those for which $S_1 = F_a^+(i+1, j+1, 0)$, $i, j \geq 1$. Then the pages of type (v) are those of the form $\mathcal{L}((F_a^+(i+1, j+1, 0) \oplus N_1)/((S_2 \cap M) \oplus N_2))$, where N_1/N_2 is a simple section of UV/F_a and M is a maximal submodule of S_1 not containing S_2. As we observed previously, each simple section of UV/F_a except $F_a^-(1,1,0)/F_a^-(0,0,1)$ is a section of $F_a^-(k+1, l+1, 0)/F_a^-(k, l, 0)$ for some $k, l \geq 1$, and clearly $S_2 \cap M = F_a^+(i, j, 0)$, for S_1 as stated. Therefore, a page of type (v) is of the form (3.11), or is a subinterval of a page (3.9). Similarly, a page of type (vi) is of the form (3.12), or is a subinterval of a page (3.9). This verifies the claim made in the previous paragraph.

The atlas of $\mathcal{L}(B/F_a)$ comprised of all pages of the forms (3.9)–(3.12) is complete (the sublattice of finite Cartesian submodules is redundant). Also, no page is a sublattice of any other page, and so we have constructed an irredundant atlas. □

The Hasse diagrams of pages (3.10)–(3.12) appear as Figures 5 and 6 in [10]. We will not need the full atlas given in Proposition 3.1.10; those pages which are relevant to our solution of the main problem are indicated in Example 3.1.11 below.

Example 3.1.11 Suppose that Q is one of the submodules cited in Lemma 3.1.9 (so that Q is a Cartesian submodule containing F_a). What are the V_4-submodules of B containing Q as a maximal submodule? That is, which edges of $\mathcal{H}(B/F_a)$ have bottom vertex Q? We will show that such edges occur in the Hasse diagram of a page (3.11), or of the page (3.10) with $i = j = k = l = 0$. These Hasse diagrams are displayed in Figure 3.1 and Figure 3.2 (without i and j set to 0), where various vertices of interest have been labelled.

Since Q is Cartesian, and a page of the atlas given in Proposition 3.1.10 is specified by a pair of Cartesian submodules, determining whether Q is contained in an atlas page is a straightforward comparison of the relevant group generating sets given by (3.2). Keeping the parameter ranges in mind, we see that Q is not contained in a page (3.9) or a page (3.12). For all values $i, j \geq 1$, the submodule $Q = F(i, j, 0, 0, 1, 1)$ is contained in pages (3.10) and (3.11); however, inspection of Figure 3.2 and then Figure 3.1 reveals that each of the three edges with bottom vertex $F(i, j, 0, 0, 1, 1)$ in the Hasse diagram of a page (3.10) is an edge in the Hasse diagram of a page (3.11). The only page containing $Q = F(0,0,0,0,1,1) = F_a$ is (3.10), with $i = j = k = l = 0$.

For all $i, j \geq 1$, $Q = F(i, j, 0, 0, 0, 1)$ is certainly contained in a page (3.11), and also in the page $\mathcal{L}(F(i, j, 1, 1, 1, 1)/F(i-1, j-1, 0, 0, 1, 1))$ of the form (3.10). Inspection of Figure 3.2 and then Figure 3.1 reveals that each of the three edges with bottom vertex $F(i, j, 0, 0, 0, 1)$ in $\mathcal{H}(F(i, j, 1, 1, 1, 1)/F(i-1, j-1, 0, 0, 1, 1))$ also occurs in the Hasse diagram of a page (3.11). By Proposition 3.1.10, we are done. □

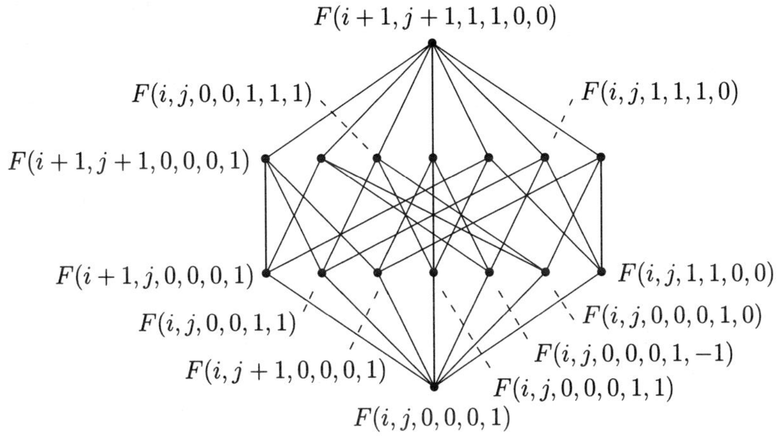

Figure 3.1: $\mathcal{H}(F(i+1,j+1,1,1,0,0)/F(i,j,0,0,0,1))$

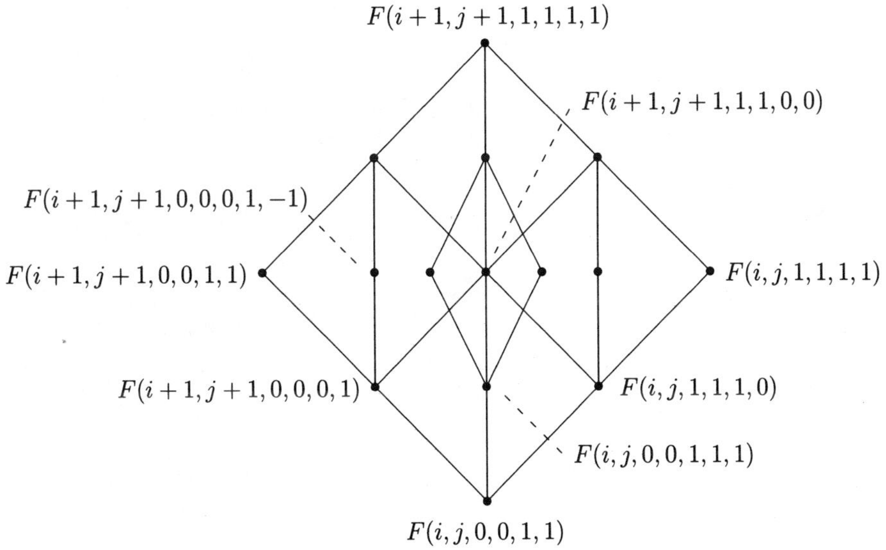

Figure 3.2: $\mathcal{H}(F(i+1,j+1,1,1,1,1)/F(i,j,0,0,1,1))$

Example 3.1.11 is needed in proof of the fundamental Theorem 3.3.10.

The concluding result of this section is used later in determining the S_4-conjugacy classes of the noncyclic finite V_4-submodules of B.

Proposition 3.1.12 *Modify the list of Theorem* 3.1.8 *by omitting*

the $F(i,j,k,l,0,0,2,\alpha)$;

the $F(i,j,k,l,0,0,\xi_1,\xi_2)$, *where* $\xi_2 \neq -1$;

the $F(i,j,k,l,0,1,1)$, *where* $k,l \geq 1$;

the $F(i,j,k,l,1,0,0)$.

A noncyclic finite V_4*-submodule of* B *is* S_4*-conjugate to at least one submodule in this modified list. If two submodules in this list are* S_4*-conjugate then they belong to the same family.*

Proof. We need to determine the S_4-orbit of each finite V_4-submodule of B containing F_a. This is known once the orbit under $\langle(1,2),(2,3)\rangle$, acting as S_3 on $\{y_j, u_j, v_j\}$, is known.

We use the shorthand of underlining various of the parameters j,k,l in a label $F(i,j,k,l,\delta,\xi,\alpha)$ to denote the set of submodules with labels obtained by permuting the underlined parameters in every possible way. For example,

$$F(i,\underline{j},k,\underline{l},\delta,\xi,\alpha) = \{F(i,j,k,l,\delta,\xi,\alpha), F(i,l,k,j,\delta,\xi,\alpha)\}.$$

A laborious yet routine calculation, using (3.2)–(3.8), shows that a submodule in the list of Theorem 3.1.8 lies in one of the following S_4-orbits:

$$\begin{aligned}
&F(i,j,\underline{k,l},0,1) \cup F(i,\underline{k},j,\underline{l},0,0,0,0) \cup F(i,\underline{l,k},j,0,0,0,1);\\
&F(i,j,\underline{k,l},1,0) \cup F(i,\underline{k},j,\underline{l},0,0,1,1) \cup F(i,\underline{l,k},j,0,0,1,0);\\
&F(i,j,\underline{k,l},1,1) \cup F(i,\underline{k},j,\underline{l},0,0,2,1,0) \cup F(i,\underline{l,k},j,0,0,2,0,1);\\
&F(i,j,\underline{k,l},0,0,1,-1) \cup F(i,\underline{k},j,\underline{l},0,0,-1,0) \cup F(i,\underline{l,k},j,0,0,-1,1);\\
&F(i,j,\underline{k,l},0,1,-1) \cup F(i,\underline{k},j,\underline{l},0,0,2,-1,0) \cup F(i,\underline{l,k},j,0,0,2,-1,1);\\
&F(i,\underline{j,k},l,1,0,1) \cup F(i,\underline{j},l,\underline{k},1,0,0) \cup F(i,l,\underline{k,j},0,1,1);\\
&F(i,j,\underline{k,l},1,0,-1) \cup F(i,\underline{k},j,\underline{l},0,0,2,1,-1) \cup F(i,\underline{l,k},j,0,0,2,0,-1);
\end{aligned}$$

$$\begin{aligned}
&F(i,\underline{j,k,l},0,0);\\
&F(i,\underline{j,k,l},0,0,-1,-1);\\
&F(i,\underline{j,k,l},0,0,0,-1);\\
&F(i,\underline{j,k,l},0,1,0);\\
&F(i,\underline{j,k,l},1,1,1);\\
&F(i,\underline{j,k,l},1,1,2,-1);\\
&F(i,\underline{j,k,l},1,1,2,1).
\end{aligned}$$

For later reference, we point out a property of submodules M in the modified list of this proposition, which is evident in the (suppressed) calculation of S_4-orbits. If M has label $F(i,j,k,l,\delta,\xi,\alpha)$ then $M^{(1,2)} = F(i,j,l,k,\delta,\xi,\alpha)$ except when $M = F(i,j,k,l,1,0,1)$; in the latter case, $M^{(1,2)} = F(i,j,l,k,1,0,0)$, which is not in the modified list.

It is apparent that some parameter values in labels in the orbit list above violate the conditions (which ensure that each listed submodule has a unique label) set down before Theorem 3.1.8. For example, in the first line, $i = j = 1$, $k = l = 0$ is admissible only in the first join component. Of course, a label "inadmissible" in this sense still defines a finite V_4-submodule of B by the relevant one of (3.2)–(3.8), and, if it contains F_a, is in the list of Theorem 3.1.8 with another label. However, we will see that this does not occur.

Apart from the $F(0,0,k,l,1,0,0)$, the labels of all submodules to be omitted have their four initial parameters positive. Inadmissible labels do not arise in the orbit of such a submodule, and we see by inspection of the orbit list above that each of them is S_4-conjugate to some submodule which is not to be omitted. We also see that $F(0,0,k,l,1,0,0)$ is S_4-conjugate to $F(0,0,l,k,1,0,1)$, the latter label being admissible and not proposed for omission. This proves the first claim of the proposition.

The second claim is equally straightforward for all cases involving only positive i,j,k,l. We calculate the additive rank of a submodule M in the modified list whose label $F(i,j,k,l,\delta,\xi,\alpha)$ has $i = j = 0$ or $k = l = 0$, and find that this rank is almost always 2 (details of the calculation are again suppressed). In such cases, $\Omega_1(M) = F_a$. If $M^r \geq F_a$ for some $r \in S_4$ then $r \in \mathsf{N}_{S_4}(F_a) = \mathsf{C}_{S_4}(\langle a \rangle) = \langle V_4, (1,2) \rangle$. Then the second claim holds for all M of rank 2, by the comments made immediately after the orbit list (at the same time, this shows that an S_4-conjugate of M containing F_a has an admissible label in the orbit list). The only M of larger rank are $F(0,0,0,0,1,1,1)$ and the $F(0,0,0,0,1,1,2,\alpha)$, all of which are normalised by S_4. □

3.2 The extension problem

In this section, we consider the problem of determining, for a transitive 2-subgroup T of S_4 and a given finite T-submodule B_1 of B, the BT-conjugacy classes of groups G such that $B \cap G = B_1$ and $BG = BT$. Our solution of this extension problem begins by calculating $|H^2(T, B_1)|$. The next step is to construct a list with the property that an extension G is BT-conjugate to at least one group in the list. If this list contains precisely $|H^2(T, B_1)|$ groups, then we have solved the extension problem for T and B_1, by Theorem 2.5. The solution method will be presented generally, before applying it in this section to the problem for $T = V_4$.

The calculation of $|H^2(T, B_1)|$ for $T = C$ is easily accomplished. In the other cases

T is dihedral, say

$$T = \langle e, f \mid e^n = f^2 = 1, e^f = e^{-1} \rangle,$$

for some $n \geq 2$. To calculate the cohomology of T, we could start with an appropriate free T-resolution of $\mathbb{Z}$ (an instance of the Wall resolution) and find $H^2(T, B_1)$ from that. This gives $H^2(T, B_1)$ as a section of $B_1^{(3)} = B_1 \oplus B_1 \oplus B_1$. However, our method must take into account practical aspects of the intended application. Determining sections of B_1 alone, for all B_1, is difficult enough to discourage us from determining the relevant sections of $B_1^{(3)}$. We could reduce the situation to one in terms of sections of B_1, using the fact that each section of a direct sum has a filtration whose quotients are sections of the summands—here we must choose from several possible filtrations. In any case, strenuous effort is required to achieve this reduction. Furthermore, the complicated nature of the resultant formula for $H^2(T, B_1)$ detracts from its usefulness as a calculational tool. An alternative approach is clearly preferable.

The approach adopted here utilises theory of the Lyndon-Hochschild-Serre (LHS) spectral sequence. The LHS spectral sequence relates cohomology of a group to cohomology of a normal subgroup (with the same coefficients) and cohomology of the corresponding quotient (with different coefficients). We consequently take advantage of the fact that T has a cyclic normal subgroup with cyclic quotient. The LHS method provides a systematically chosen filtration that allows us to calculate directly in terms of sections of B_1. The calculation breaks down into several parts, each of which is reasonably straightforward to undertake with the B_1 we have to hand.

Familiarity with basic concepts and notation in the theory of spectral sequences, as found in Chapter XI of [15], will be assumed. The LHS (cohomology) spectral sequence will be denoted here as $\{E_r, d_r\}$, with terms $E_r^{p,q}$ ($= 0$ if $p < 0$ or $q < 0$) and differentials $d_r^{p,q}\colon E_r^{p,q} \to E_r^{p+r,q-r+1}$. As part of the definition of spectral sequence, $d_r^{p-r,q+r-1} d_r^{p,q} = 0$ and $E_{r+1}^{p,q} \cong \ker d_r^{p,q} / \operatorname{im} d_r^{p-r,q+r-1}$. We define the LHS spectral sequence in particular relative to T and the normal subgroup $N = \langle e \rangle$ of T. This spectral sequence has the following properties:

(i) it converges to the graded module $H^{p+q}(T, B_1)$,

(ii) $E_2^{p,q} \cong H^p(T/N, H^q(N, B_1))$.

Convergence is defined via some filtration of the graded module $H^{p+q}(T, B_1)$, and in total degree $p + q = 2$ yields the formula

$$|H^2(T, B_1)| = |\operatorname{coker} d_2^{0,1}| |\ker d_2^{1,1}| |\ker d_3^{0,2}|. \tag{3.13}$$

Thus, our calculation of $|H^2(T, B_1)|$ depends on determining the action of $d_2^{0,1}, d_2^{1,1}$ and $d_3^{0,2}$. To do this we will construct the LHS spectral sequence from first principles.

Let M be an arbitrary (right) T-module. The endomorphism of M induced by action of an element of $\mathbb{Z}T$ will be written conveniently as that element. Set $\hat{e} = \sum_{i=0}^{n-1} e^i$. By virtue of identities in $\mathbb{Z}T$ such as

$$\begin{aligned} (1-f)(1-e) &= (1-e)(1+fe) \\ (1+fe)\hat{e} &= \hat{e}(1+f) \\ (1+f)(1-e) &= (1-e)(1-fe) \\ (1-fe)\hat{e} &= \hat{e}(1-f), \end{aligned}$$

we may write down a double $\mathbb{Z}T$-complex whose terms are $\mathbb{Z}T_{\mathbb{Z}T}$, vertical differentials are $\pm\hat{e}, \pm(1-e)$, and horizontal differentials are $1 \pm f, 1 \pm fe$. These differentials act by multiplication on the left in $\mathbb{Z}T$. Each row and each column of the double complex is exact except at the ultimate terms. We now apply to this double complex the contravariant functor $\mathrm{Hom}_{\mathbb{Z}T}(-, M)$. Since $\mathrm{Hom}_{\mathbb{Z}T}(\mathbb{Z}T, M) \cong M$ as right T-modules, we obtain the double $\mathbb{Z}$-cocomplex

$$\begin{array}{ccccccccc}
\vdots & & \vdots & & \vdots & & \vdots & & \\
{\scriptstyle \hat{e}}\uparrow & & {\scriptstyle -\hat{e}}\uparrow & & {\scriptstyle \hat{e}}\uparrow & & {\scriptstyle -\hat{e}}\uparrow & & \\
M & \xrightarrow{1-fe} & M & \xrightarrow{1+fe} & M & \xrightarrow{1-fe} & M & \longrightarrow & \cdots \\
{\scriptstyle 1-e}\uparrow & & {\scriptstyle -(1-e)}\uparrow & & {\scriptstyle 1-e}\uparrow & & {\scriptstyle -(1-e)}\uparrow & & \\
M & \xrightarrow{1+f} & M & \xrightarrow{1-f} & M & \xrightarrow{1+f} & M & \longrightarrow & \cdots \\
{\scriptstyle \hat{e}}\uparrow & & {\scriptstyle -\hat{e}}\uparrow & & {\scriptstyle \hat{e}}\uparrow & & {\scriptstyle -\hat{e}}\uparrow & & \\
M & \xrightarrow{1+fe} & M & \xrightarrow{1-fe} & M & \xrightarrow{1+fe} & M & \longrightarrow & \cdots \\
{\scriptstyle 1-e}\uparrow & & {\scriptstyle -(1-e)}\uparrow & & {\scriptstyle 1-e}\uparrow & & {\scriptstyle -(1-e)}\uparrow & & \\
M & \xrightarrow{1-f} & M & \xrightarrow{1+f} & M & \xrightarrow{1-f} & M & \longrightarrow & \cdots
\end{array}$$

whose terms will be denoted $E_0^{p,q}$, viewing this as the E_0 page (not the lattice-theoretic term) of the spectral sequence to be constructed. Denote the horizontal differentials $\partial_{\rightarrow}^{p,q}$ and the vertical differentials $\partial_{\uparrow}^{p,q}$. We now begin construction of the LHS spectral sequence by the recipe outlined in §8, Chapter XI of [15].

First, we form the total cocomplex A of the double cocomplex, whose terms are the $A^n = \bigoplus_{p+q=n} E_0^{p,q}$. The filtration $\cdots \subseteq F^{p+1}A \subseteq F^pA \subseteq F^{p-1}A \subseteq \cdots$ of A is defined as usual by

$$F^pA^n = \bigoplus_{k \geq p} E_0^{k,n-k}.$$

There is a differential $\partial\colon F^pA^n \to F^pA^{n+1}$ induced by the differential $\partial = \partial_{\rightarrow} + \partial_{\uparrow}$ on A. Set

$$Z_r^{p,q} = \{m \in F^pA^{p+q} \mid m\partial \in F^{p+r}A^{p+q+1}\}.$$

Then $E_r^{p,q}$ is defined by

$$E_r^{p,q} = \frac{Z_r^{p,q} + F^{p+1}A^{p+q}}{Z_{r-1}^{p-r+1,q+r-2}\partial + F^{p+1}A^{p+q}}$$

and $d_r^{p,q}$ is the differential induced on this section of F^pA^{p+q} by ∂. For example, we have the following explicit definition of $\partial\colon F^pA^1 \to F^pA^2$ for $p \leq 0$:

$$(m,m') \in E_0^{0,1} \oplus E_0^{1,0} \mapsto (m\partial_\uparrow^{0,1}, m\partial_\rightarrow^{0,1} + m'\partial_\uparrow^{1,0}, m'\partial_\rightarrow^{1,0}) \in E_0^{0,2} \oplus E_0^{1,1} \oplus E_0^{2,0}.$$

We assert that the total complex $\tilde{A}$ of the original double complex is a free resolution of $\mathbb{Z}$ as a $\mathbb{Z}T$-module. One way to see this is to form the (homology) spectral sequence of the double complex, by evaluating first vertical and then horizontal homology. The E^1 page of this spectral sequence is zero everywhere except along the bottom row, where it forms a $\mathbb{Z}(T/\langle e\rangle)$-resolution of $\mathbb{Z}T/(1-e)\mathbb{Z}T$. The E^2 page has $\mathbb{Z}$ in the $(0,0)$ position and 0 elsewhere. But this spectral sequence converges to the homology of $\tilde{A}$ (see Theorem 6.1, p.341 of [15]), verifying the assertion. Then by Theorem 3.1, p.327 of [15], the cohomology spectral sequence $\{E_r, d_r\}$ defined above converges to

$$H_{p+q}A = H_{p+q}(\mathrm{Hom}_{\mathbb{Z}T}(\tilde{A}, M)) = H^{p+q}(T, M).$$

Furthermore, since $E_1^{p,q} \cong \ker \partial_\uparrow^{p,q}/\mathrm{im}\,\partial_\uparrow^{p,q-1}$ and $d_1^{p,q} = \partial_\rightarrow^{p,q}$, it is clear that $E_2^{p,q} \cong H^p(T/N, H^q(N,M))$. Thus, we have indeed constructed the LHS spectral sequence for this T and N.

In the following suite of results we determine explicitly the action of the differentials appearing in (3.13). The proofs are elementary, similar, and omitted: they proceed directly from the definitions and the discussion above. To simplify notation, cosets in a submodule of M (clear from the context) are denoted by $[-]$. Also, the superscript -1 on an element of $\mathbb{Z}T$ denotes the complete inverse image in M of the endomorphism induced by that element, and "ker" means kernel of a homomorphism on M.

Proposition 3.2.1 *In the notation above,*

$$E_2^{0,1} = (\ker\hat{e} \cap M(1-e)(1+fe)^{-1})/M(1-e),$$
$$E_2^{2,0} = (\ker(1-e) \cap \ker(1-f))/(\ker(1-e))(1+f).$$

For $m \in \ker\hat{e} \cap (M(1-e))(1+fe)^{-1}$, choose $m' \in M$ such that

$$m(1+fe) = m'(1-e);$$

then $d_2^{0,1}\colon E_2^{0,1} \to E_2^{2,0}$ is defined by

$$[m] \mapsto [m'(1+f)].$$

Proposition 3.2.2 *In the notation above,*

$$E_2^{1,1} = (\ker\hat{e} \cap M(1-e)(1-fe)^{-1})/((\ker\hat{e})(1+fe) + M(1-e)),$$
$$E_2^{3,0} = (\ker(1-e) \cap \ker(1+f))/(\ker(1-e))(1-f).$$

For $m \in \ker \hat{e} \cap M(1-e)(1-fe)^{-1}$, *choose* $m' \in M$ *such that*

$$m(1-fe) = -m'(1-e);$$

then $d_2^{1,1}\colon E_2^{1,1} \to E_2^{3,0}$ *is defined by*

$$[m] \mapsto [m'(1-f)].$$

Proposition 3.2.3 *In the notation above,*

$$\begin{aligned} E_3^{0,2} &= (\ker(1-e) \cap M(1-e)(1-fe)^{-1}\hat{e}(1+f)^{-1})/M\hat{e}, \\ E_3^{3,0} &= (\ker(1-e) \cap \ker(1+f))/(\ker \hat{e})(1-fe)(1-e)^{-1}(1-f). \end{aligned}$$

For $m \in \ker(1-e) \cap M(1-e)(1-fe)^{-1}\hat{e}(1+f)^{-1}$, *choose* $m' \in M$ *and* $m'' \in M$ *such that*

$$\begin{aligned} m(1+f) &= m'\hat{e}, \\ m'(1-fe) &= -m''(1-e); \end{aligned}$$

then $d_3^{0,2}\colon E_3^{0,2} \to E_3^{3,0}$ *is defined by*

$$[m] \mapsto [m''(1-f)].$$

We are now in a position to calculate $|H^2(V_4, B_1)|$, for which we replace e by a, f by b, and M by B_1 in the preceding results. The next lemma is readily verified.

Lemma 3.2.4 *Assume the notation of Lemma* 3.1.6 *and let* s, t *be distinct involutions of* V_4. *Suppose that* $M \le C_s$ *or* $M \le X$. *Then*

$$(B_1 \cap MC_t)^{1-s} = \begin{cases} (B_1 \cap C_t)^2 & \textit{if } B_1 \cap MC_t = (B_1 \cap M)(B_1 \cap C_t) \\ B_1 \cap C_t & \textit{otherwise} \end{cases}$$

and

$$(B_1 \cap MC_t)^{1+s} = \begin{cases} (B_1 \cap M)^2 & \textit{if } B_1 \cap MC_t = (B_1 \cap M)(B_1 \cap C_t) \\ B_1 \cap M & \textit{otherwise.} \end{cases}$$

Lemmas 3.1.6 and 3.2.4 will be used implicitly and repeatedly in the subsequent calculations. These mainly involve comparing intersections of B_1 and products of the X, Y, U, V. For example, $\ker(1-t) = B_1 \cap XC_t$ and $\ker(1+t) = B_1 \cap C_sC_{s'}$, where s, s' and t are distinct involutions of V_4. Then $E_2^{2,0} = (B_1 \cap X)/(B_1 \cap XY)^{1+b}$ and $E_2^{3,0} = (B_1 \cap Y)/(B_1 \cap XY)^{1-b}$, by Propositions 3.2.1 and 3.2.2. By Lemma 3.2.4, this shows that

$$|E_2^{2,0}| = |E_2^{3,0}| = \begin{cases} 2 & \text{if } B_1 \cap XY = (B_1 \cap X)(B_1 \cap Y) \\ 1 & \text{otherwise.} \end{cases} \tag{3.14}$$

Lemma 3.2.5 *In either of the following situations,* $|\mathrm{coker}\, d_2^{0,1}| = 1$:

(i) $B_1 \cap XY \neq (B_1 \cap X)(B_1 \cap Y)$,

(ii) $B_1 \cap XY = (B_1 \cap X)(B_1 \cap Y)$ *and* $B_1 \cap UV \neq (B_1 \cap U)(B_1 \cap V)$ *and* $B_1 \cap XYV \neq (B_1 \cap X)(B_1 \cap YV)$.

In all other situations, $|\mathrm{coker}\, d_2^{0,1}| = 2$.

Proof. If (i) occurs then certainly $|\mathrm{coker}\, d_2^{0,1}| = 1$, since $d_2^{0,1}$ has trivial codomain by (3.14). For the rest of the proof assume that $B_1 \cap XY = (B_1 \cap X)(B_1 \cap Y)$, so $|E_2^{2,0}| = 2$ by (3.14). Choose $m \in \ker(1+a) = B_1 \cap UV$ and $m' \in B_1$ such that $m^{1+ab} = (m')^{1-a}$. Write $m = uv$ for some $u \in U$ and $v \in V$. Then $(m')^{1-a} = v^{1-a}$, implying that $m' = xy\overline{v}$ for some $x \in X$, $y \in Y$, $\overline{v} \in V$. By Proposition 3.2.1, $(mB_1^{1-a})d_2^{0,1} = x^2(B_1 \cap X)^2$. If $B_1 \cap UV = (B_1 \cap U)(B_1 \cap V)$ then we may choose $m' = v$. If $B_1 \cap XYV = (B_1 \cap X)(B_1 \cap YV)$ then $x \in B_1$. In either case, $|\mathrm{coker}\, d_2^{0,1}| = 2$. Finally, suppose $B_1 \cap UV \neq (B_1 \cap U)(B_1 \cap V)$ and $B_1 \cap XYV \neq (B_1 \cap X)(B_1 \cap YV)$, so that we may choose $uv \in B_1$ with $v \notin B_1$, and $xy\overline{v} \in B_1$ with $x \notin B_1$, $y\overline{v} \notin B_1$. Then $\overline{v} \notin B_1$ and $v^2(B_1 \cap V)^2 = \overline{v}^2(B_1 \cap V)^2$. So for $m = uv$ there is $m' \in B_1$ such that $m^{1+ab} = (m')^{1-a}$ and $m'YV \in xYV$; as a consequence, $\mathrm{im}\, d_2^{0,1} \neq 1$. □

The next lemma is an application of Proposition 3.2.2, and may be proved in the same way that Lemma 3.2.5 was proved.

Lemma 3.2.6 *If one of the following situations occurs then* $\ker d_2^{1,1} = E_2^{1,1}$:

(i) $B_1 \cap XY \neq (B_1 \cap X)(B_1 \cap Y)$,

(ii) $B_1 \cap UV = (B_1 \cap U)(B_1 \cap V)$,

(iii) $B_1 \cap XY = (B_1 \cap X)(B_1 \cap Y)$ *and* $B_1 \cap UV \neq (B_1 \cap U)(B_1 \cap V)$ *and* $B_1 \cap XYU = (B_1 \cap XU)(B_1 \cap Y)$.

In all other situations, $|\ker d_2^{1,1}| = |E_2^{1,1}|/2$.

Lemma 3.2.7 $\ker d_3^{0,2} = E_3^{0,2}$.

Proof. We compare $B_1 \cap XY$ and $(B_1 \cap X)(B_1 \cap Y)$. If these do not coincide then $E_3^{3,0}$ (as the section $\ker d_2^{3,0}/\mathrm{im} d_2^{1,1}$ of $E_2^{3,0}$) is trivial by (3.14), so that $d_3^{0,2}$ is trivial. On the other hand, suppose that $B_1 \cap XY = (B_1 \cap X)(B_1 \cap Y)$. In Proposition 3.2.3, $m = xy$ for some $x \in B_1 \cap X$, $y \in B_1 \cap Y$, and we choose $m' = x$ and $m'' = 1$. By the proposition, $d_3^{0,2}$ is trivial in this case also. □

From now on, we assume that $B_1 = F(i,j,k,l,\delta,\xi,\alpha)$ is in the list of Theorem 3.1.8. The orders in (3.13) may be calculated by Propositions 3.2.1–3.2.3 and Lemmas 3.2.5–3.2.7. Note that comparison of $B_1 \cap XY$ with $(B_1 \cap X)(B_1 \cap Y)$, or of $B_1 \cap UV$ with $(B_1 \cap U)(B_1 \cap V)$, is made in the maximal Cartesian submodule $F(i,j,k,l,\delta)$ of B_1.

By Lemma 3.2.5, if $\delta_1 = 1$ then $|\operatorname{coker} d_2^{0,1}| = 1$, and if $\delta_1 = \delta_2 = 0$ then $|\operatorname{coker} d_2^{0,1}| = 2$. If $\delta_1 = 0$ and $\delta_2 = 1$ then $|\operatorname{coker} d_2^{0,1}| = 2$ if $\xi = 0$ or is a void, and $|\operatorname{coker} d_2^{0,1}| = 1$ if $\xi \neq 0$.

In determining $|E_2^{1,1}|$, we consider separately the cases $\delta_2 = 0$ and $\delta_2 = 1$. Suppose that $\delta_2 = 0$. Then $E_2^{1,1} = (B_1 \cap UV)/B_1^{1-a}$ by Proposition 3.2.2; that is, $|E_2^{1,1}|$ depends on the index of $F(i,j,k,l,\delta)$ in B_1, which may be read off the label of B_1. Specifically, $|E_2^{1,1}| = 4, 2$ or 1 when the index is $1, 2$ or 4 respectively; in terms of the label, ξ is a void, α is a void and ξ is not, or $\xi = 2$, respectively. Now suppose that $\delta_2 = 1$. Then $B_1^{1-a} = B_1 \cap UV$ if the index of $F(i,j,k,l,\delta)$ in B_1 is 4, whence $|E_2^{1,1}| = 1$ by Proposition 3.2.2. Similarly, $|E_2^{1,1}| = 1$ if ξ is a void, and $|E_2^{1,1}| = 2$ if α but not ξ is a void.

We determine $|E_3^{0,2}|$ by Proposition 3.2.3 and reasoning similar to that used in the previous paragraph. If $\delta_1 = 0$ then $|E_3^{0,2}| = 4, 2$ or 1 according to whether ξ is a void, α but not ξ is a void, or $\xi = 2$, respectively. Now suppose that $\delta_1 = 1$. If $\xi = 2$ then $|E_3^{0,2}| = 1$. If α but not ξ is a void then $|E_3^{0,2}| = 2$ precisely when either $B_1 \cap XU \neq (B_1 \cap X)(B_1 \cap U)$ or $B_1 \cap XV \neq (B_1 \cap X)(B_1 \cap V)$; otherwise, $|E_3^{0,2}| = 1$. If ξ is a void then $|E_3^{0,2}| = 2$.

Together with Lemmas 3.2.6, 3.2.7 and (3.13), the results derived in the previous three paragraphs are used to compile Table 3.1 below.

<table>
<tr><th>B_1</th><th>$|\operatorname{coker} d_2^{0,1}|$</th><th>$|\ker d_2^{1,1}|$</th><th>$|\ker d_3^{0,2}|$</th><th>$|H^2(V_4, B_1)|$</th></tr>
<tr><td>$F(i,j,k,l,0,0)$</td><td rowspan="4">2</td><td>4</td><td>4</td><td>32</td></tr>
<tr><td>$F(i,j,k,l,0,0,\xi)$</td><td>2</td><td>2</td><td>8</td></tr>
<tr><td>$F(i,j,k,l,0,0,2,\alpha)$</td><td rowspan="2">1</td><td>1</td><td>2</td></tr>
<tr><td>$F(i,j,k,l,0,1)$</td><td>4</td><td>8</td></tr>
<tr><td>$F(i,j,k,l,0,1,1)$</td><td rowspan="2">1</td><td>2</td><td rowspan="5">2</td><td>4</td></tr>
<tr><td>$F(i,j,k,l,0,1,-1)$</td><td rowspan="2">1</td><td>2</td></tr>
<tr><td>$F(i,j,k,l,0,1,0)$</td><td>2</td><td>4</td></tr>
<tr><td>$F(i,j,k,l,1,0)$</td><td rowspan="7">1</td><td>4</td><td>8</td></tr>
<tr><td>$F(i,j,k,l,1,0,1)$</td><td rowspan="3">2</td><td>4</td></tr>
<tr><td>$F(i,j,k,l,1,0,-1)$</td><td>1</td><td>2</td></tr>
<tr><td>$F(i,j,k,l,1,0,0)$</td><td rowspan="3">2</td><td>4</td></tr>
<tr><td>$F(i,j,k,l,1,1)$</td><td>1</td><td>2</td></tr>
<tr><td>$F(i,j,k,l,1,1,1)$</td><td>2</td><td>4</td></tr>
<tr><td>$F(i,j,k,l,1,1,2,\alpha)$</td><td>1</td><td>1</td><td>1</td></tr>
</table>

Table 3.1: Calculation of $|H^2(V_4, B_1)|$

In the sequel, $\sim$ denotes conjugacy. Some allied notation is: $\sim_z$, to make explicit the conjugating element z; and $\sim_S$, for conjugacy by an element of $S \subseteq GL(4)$.

We move on now to construction of V_4-extensions in BV_4 (at this stage we could, but do not, enforce irreducibility). The next lemma deals with the splitting case.

Lemma 3.2.8 *Let G be a finite subgroup of BS_4 such that $BG = BT$ for some regular subgroup T of S_4. If G splits over $B \cap G$ then $G \sim_{BT} (B \cap G) \rtimes T$.*

Proof. If K is a complement of $B \cap G$ in G then K is a complement of B in $B \rtimes T$. Since T is itself such a complement and $H^1(T, B) = 0$ by Lemma 2.2, the result is clear. $\square$

Theorem 3.2.9 *A finite subgroup G of BV_4 such that $BG = BV_4$ and $B \cap G$ is noncyclic is BS_4-conjugate to one of the groups in the following list:*

$$\begin{array}{l}
\langle ax_{i+1}^{\varepsilon}y_{j+1}^{\eta}, bx_{i+1}^{\gamma}u_{k+1}^{\mu}v_{l+1}^{\nu}, F(i,j,k,l,0,0)\rangle, \\
\langle ax_{i+1}^{\varepsilon}, bx_{i+1}^{\gamma}u_{k+1}^{\mu}, F(i,j,k,l,0,0,-1,-1)\rangle, \\
\langle ax_{i+1}^{\varepsilon}, bx_{i+1}^{\gamma}u_{k+1}^{\mu}, F(i,j,k,l,0,0,0,-1)\rangle, \\
\langle ay_{j+1}^{\eta}, bx_{i+1}^{\gamma}u_{k+1}^{\mu}, F(i,j,k,l,0,0,1,-1)\rangle, \\
\langle ax_{i+1}^{\varepsilon}y_{j+1}^{\eta}, bx_{i+1}^{\gamma}, F(i,j,k,l,0,1)\rangle, \\
\langle ay_{j+1}^{\eta}, b(x_{i+2}u_2v_2)^{\gamma}, F(i,j,0,0,0,1,1)\rangle, \\
\langle ax_{i+1}^{\varepsilon}, b, F(i,j,k,l,0,1,-1)\rangle, \\
\langle ax_{i+1}^{\varepsilon}, bx_{i+1}^{\gamma}, F(i,j,k,l,0,1,0)\rangle, \\
\langle ax_{i+1}^{\varepsilon}, bu_{k+1}^{\mu}v_{l+1}^{\nu}, F(i,j,k,l,1,0)\rangle, \\
\langle a(x_{i+2}y_{j+2})^{\varepsilon}, b(x_{i+2}u_{k+2})^{\varepsilon}v_{l+1}^{\nu}, F(i,j,k,l,1,0,1)\rangle, \\
\langle a, bv_{l+1}^{\nu}, F(i,j,k,l,1,0,-1)\rangle, \\
\langle ax_{i+1}^{\varepsilon}, b, F(i,j,k,l,1,1)\rangle, \\
\langle a(x_{i+2}y_{j+2})^{\varepsilon}, b(x_{i+2}u_{k+2})^{\gamma}v_{l+2}^{\varepsilon-\gamma}, F(i,j,k,l,1,1,1)\rangle, \\
\langle a, b, F(i,j,k,l,1,1,2,-1)\rangle, \\
\langle a, b, F(i,j,k,l,1,1,2,1)\rangle,
\end{array}$$

where $\varepsilon, \eta, \gamma, \mu, \nu$ range over $\{0,1\}$ and i,j,k,l range over the non-negative integers under the conditions summarised after Theorem 3.1.8. Distinct groups in this list are not BV_4-conjugate.

Proof. Set $B \cap G = B_1$. There is a B-conjugate of G of the form

$$\langle ax_ay, bx_buv, B_1\rangle$$

where $x_a, x_b \in X$, $y \in Y$, $u \in U$ and $v \in V$. For certainly $G = \langle ax_ay\tilde{u}\tilde{v}, bx_b\tilde{y}uv, B_1\rangle$ where $\tilde{y} \in Y$, $\tilde{u} \in U$, $\tilde{v} \in V$, and if we choose $\hat{y} \in Y$, $\hat{u} \in U$, $\hat{v} \in V$ such that

$\hat{y}^2 = \tilde{y}^{-1}$, $\hat{u}^2 = \tilde{u}^{-1}$, $\hat{v}^2 = \tilde{v}^{-1}$, then $G^{\hat{y}\hat{u}\hat{v}} = \langle ax_a y, bx_b uv\tilde{v}^{-1}, B_1\rangle$: this justifies the claim. The relations $a^2 = b^2 = 1$ in $V_4 \cong G/B_1$ imply that

$$x_a^2 y^2 \in B_1 \cap XY, \tag{3.15}$$

$$x_b^2 u^2 \in B_1 \cap XU. \tag{3.16}$$

The relation $[a, b] = 1$ in V_4 implies that $y^{-2}u^2v^2 \in B_1 \cap YUV$. However, $y^4 \in B_1 \cap Y$ by (3.15), since $(B_1 \cap X)(B_1 \cap Y)$ has index at most 2 in $B_1 \cap XY$. Therefore, we write the last condition more conveniently as

$$y^2u^2v^2 \in B_1 \cap YUV. \tag{3.17}$$

Conversely, any group generated by B_1, $ax_a y$, $bx_b uv$ and satisfying (3.15)–(3.17) is a finite supplement of B in BV_4 with diagonal subgroup B_1.

For each choice of B_1, we must show that G is BS_4-conjugate to at least one of $|H^2(V_4, B_1)|$ groups constructed to satisfy the criteria stated in the previous paragraph. Once this is done, and since BV_4-conjugate groups in BV_4 have identical diagonal subgroups, the second claim of the theorem is guaranteed by Theorem 2.5.

Clearly, we may assume that B_1 is in the list of Proposition 3.1.12. Suppose that B_1 has label $F(i, j, k, l, \delta, \xi, \alpha)$. When $\delta = (1, 1)$ and $\xi = 2$, G is BV_4-conjugate to the split extension $\langle a, b, B_1\rangle$, by Lemma 3.2.8 and Table 3.1. The manipulations required in the non-splitting cases are less straightforward. They are, however, repetitive in nature, and will be illustrated with only a few typical examples.

In the rest of the proof, $g_1 \equiv g_2$ for $g_1, g_2 \in G$ means that $g_1 g_2^{-1} \in B_1$; equality modulo any other subgroup S of B will be written $g_1 \equiv g_2 \bmod S$.

Initially suppose that $\delta = (0, 0)$. Then $x_a^2 \in B_1 \cap X = \langle x_i\rangle$ and $y^2 \in B_1 \cap Y = \langle y_j\rangle$ by (3.15). That is, $x_a \equiv 1$ or x_{i+1} and $y \equiv 1$ or y_{j+1}, so that of course we may take $x_a = 1$ or x_{i+1} and $y = 1$ or y_{j+1} in the generating set for G. Similarly, by (3.17) we may take $u = 1$ or u_{k+1} and $v = 1$ or v_{l+1}; then $x_b = 1$ or x_{i+1} by (3.16). For fixed i, j, k, l, there are thus 32 distinct extensions of B_1 by V_4 in BV_4. If B_1 is Cartesian then each extension appears in the list of this theorem, by Table 3.1, and we are done. If B_1 is non-Cartesian and $\xi = (1, -1)$ then

$$\begin{aligned} \langle ax_{i+1}y, bx_b uv, B_1\rangle &= \langle ayu_{k+1}v_{l+1}, bx_b uv, B_1\rangle \\ &\sim_{u_{k+2}v_{l+2}} \langle ay, bx_b uvv_{l+1}, B_1\rangle. \end{aligned}$$

Furthermore, since $v_{l+1} \equiv x_{i+1}u_{k+1}$ here,

$$\langle ay, bx_b uv_{l+1}, B_1\rangle = \langle ay, bx_b x_{i+1} uu_{k+1}, B_1\rangle.$$

This shows that G is B-conjugate to one of the eight groups arising as η, γ, μ range (independently) over $\{0, 1\}$ in $\langle ay_{j+1}^{\eta}, bx_{i+1}^{\gamma}u_{k+1}^{\mu}, B_1\rangle$. By Table 3.1, all of these groups are listed.

Next suppose that $\delta = (0,1)$: then we have the same possibilities for x_a and y as before. If $\xi = -1$ then by (3.16) and (3.17) we also have the same possibilities for x_b, u and v as before. However, $y_{j+1}u_{k+1} \equiv x_{i+1}$ in this case, so that

$$\langle ax_a y_{j+1}, bx_b uv, B_1\rangle \sim_{u_{k+2}} \langle ax_a x_{i+1}, bx_b uv, B_1\rangle$$

and

$$\langle ax_a, bx_{i+1}uv, B_1\rangle \sim_{y_{j+2}} \langle ax_a, buu_{k+1}v, B_1\rangle.$$

We also have $u_{k+1} \equiv v_{l+1}$ and $\langle ax_a, bv_{l+1}, B_1\rangle \sim_{u_{k+2}v_{l+2}} \langle ax_a, b, B_1\rangle$. All of this implies that G is B-conjugate to $\langle a, b, B_1\rangle$ or $\langle ax_{i+1}, b, B_1\rangle$. Both groups are listed, by Table 3.1.

As the final example, suppose that $\delta = (1,0)$ and $\xi = 1$. By (3.15), $x_a y \equiv 1$ or $x_{i+2}y_{j+2}$ mod $\langle x_{i+1}, y_{j+1}\rangle$, and by (3.16), $x_b u \equiv 1$ or $x_{i+2}u_{k+2}$ mod $\langle x_{i+1}, u_{k+1}\rangle$. Since $x_{i+1} \equiv y_{j+1} \equiv u_{k+1}$, after possible conjugation by u_{k+2} or y_{j+2} we may take $x_a y = 1$ or $x_{i+2}y_{j+2}$, and $x_b u = 1$ or $x_{i+2}u_{k+2}$. When $x_b u = x_{i+2}u_{k+2}$, necessarily $x_a y = x_{i+2}y_{j+2}$ and $v = 1$ or v_{l+1} by (3.17). When $x_b u = 1$, necessarily $x_a y = 1$ and $v = 1$ or v_{l+1}. Given these options, we see that there are $|H^2(V_4, B_1)| = 4$ extensions of B_1 by V_4 in BV_4, all of which are listed. □

3.3 The conjugacy problem

In this section we will produce a complete list of the finite irreducible subgroups of BV_4. Our method is as follows. First, in Theorem 3.3.1, we modify the list in Theorem 3.2.9 by omitting reducible groups and all but one representative from each BS_4-conjugacy class. This is done in such a way a way that distinct diagonal subgroups of representatives chosen are not S_4-conjugate (see Remark 3.3.2).

If two distinct groups in the list of Theorem 3.3.1 are $GL(4)$-conjugate then a conjugating element cannot map diagonal subgroup onto diagonal subgroup: if it did, then by Remark 3.3.2 and Theorem 2.7 the diagonal subgroups would coincide. The conjugating element would be monomial by Proposition 1.3.7, and the two groups would be BS_4-conjugate by Remark 1.3.8. This suggests that we should distinguish between those groups in the list of Theorem 3.3.1 which have a unique self-centralising normal subgroup with quotient isomorphic to V_4, and those which do not. (Henceforth, we refer to such a subgroup by the abbreviation SCN4; of course, the diagonal subgroup at least is SCN4.) We have just seen that distinct groups of the former kind are not conjugate. Therefore, the final task is to determine all conjugacy between groups of the latter kind.

Theorem 3.3.1 *The distinct groups that are defined by the generating sets below, where $\varepsilon, \eta, \gamma, \mu, \nu$ range independently over $\{0,1\}$ and i,j,k,l range over the non-negative integers as indicated, constitute a complete and irredundant list of BS_4-conjugacy class representatives of the finite irreducible subgroups of BV_4 with noncyclic diagonal subgroup.*

For $i,j,k,l \geq 1$:

$$\begin{array}{l}
\langle ax_{i+1}^{\varepsilon}y_{j+1}^{\eta}, bx_{i+1}^{\gamma}u_{k+1}^{\mu}v_{l+1}^{\nu}, F(i,j,k,l,0,0)\rangle, \quad j<k<l; \\
\langle ay_{j+1}^{\eta}, bx_{i+1}^{\gamma}u_{k+1}^{\mu}v_{k+1}^{\nu\mu}, F(i,j,k,k,0,0)\rangle, \quad j \neq k; \\
\langle ax_{i+1}y_{j+1}^{\eta}, bu_{k+1}^{\mu}v_{k+1}^{\nu}, F(i,j,k,k,0,0)\rangle, \quad j \neq k; \\
\langle ay_{j+1}^{\mu}, bu_{j+1}^{\mu}v_{j+1}^{\nu(1-\mu)+1}, F(i,j,j,j,0,0)\rangle; \\
\langle ax_{i+1}y_{j+1}^{\eta}, bu_{j+1}^{\mu}v_{j+1}^{\nu\eta}, F(i,j,j,j,0,0)\rangle; \\
\langle ay_{j+1}^{\eta}, bx_{i+1}^{\gamma}u_{k+1}^{\mu}, F(i,j,k,l,0,0,1,-1)\rangle, \quad k<l; \\
\langle ay_{j+1}^{\eta}, bx_{i+1}^{\gamma(1-\mu)}u_{k+1}^{\mu}, F(i,j,k,k,0,0,1,-1)\rangle; \\
\langle ax_{i+1}^{\varepsilon}, bx_{i+1}^{\gamma}u_{k+1}^{\mu}, F(i,j,k,l,0,0,0,-1)\rangle, \quad j<k<l; \\
\langle ax_{i+1}^{\varepsilon}, bx_{i+1}^{\gamma(1-\varepsilon)}u_{k+1}^{\mu}, F(i,j,k,k,0,0,0,-1)\rangle, \quad j \neq k; \\
\langle ax_{i+1}^{\varepsilon}, bu_{j+1}^{\mu}, F(i,j,j,j,0,0,0,-1)\rangle; \\
\langle ax_{i+1}^{\varepsilon}, bx_{i+1}^{\gamma}u_{k+1}^{\mu}, F(i,j,k,l,0,0,-1,-1)\rangle, \quad j<k<l; \\
\langle ax_{i+1}^{\varepsilon}, bx_{i+1}^{\gamma(1-\varepsilon-\mu)}u_{k+1}^{\mu}, F(i,j,k,k,0,0,-1,-1)\rangle, \quad j \neq k; \\
\langle ax_{i+1}^{\varepsilon}, bu_{j+1}^{\mu}, F(i,j,j,j,0,0,-1,-1)\rangle;
\end{array}$$

for $i,j \geq 1$ and $k,l \gtrsim 0$:

$$\begin{array}{l}
\langle ax_{i+1}^{\varepsilon}y_{j+1}^{\eta}, bx_{i+1}^{\gamma}, F(i,j,k,l,0,1)\rangle, \quad k<l; \\
\langle ax_{i+1}^{\varepsilon}y_{j+1}^{\eta}, bx_{i+1}^{\gamma(1-\varepsilon)}, F(i,j,k,k,0,1)\rangle; \\
\langle ay_{j+1}^{\eta}, b(x_{i+2}u_2v_2)^{\gamma}, F(i,j,0,0,0,1,1)\rangle; \\
\langle ax_{i+1}^{\varepsilon}, b, F(i,j,k,l,0,1,-1)\rangle, \quad k \leq l; \\
\langle ax_{i+1}^{\varepsilon}, bx_{i+1}^{\gamma}, F(i,j,k,l,0,1,0)\rangle, \quad j<k<l; \\
\langle ax_{i+1}^{\varepsilon}, bx_{i+1}^{\gamma(1-\varepsilon)}, F(i,j,k,k,0,1,0)\rangle, \quad j \neq k; \\
\langle ax_{i+1}^{\varepsilon}, b, F(i,j,j,j,0,1,0)\rangle;
\end{array}$$

for $i,j \gtrsim 0$ and $k,l \geq 1$:

$$\begin{array}{l}
\langle ax_{i+1}^{\varepsilon}, bu_{k+1}^{\mu}v_{l+1}^{\nu}, F(i,j,k,l,1,0)\rangle, \quad k<l; \\
\langle ax_{i+1}^{\varepsilon}, bu_{k+1}^{\mu}v_{k+1}^{\nu\mu}, F(i,j,k,k,1,0)\rangle; \\
\langle a(x_{i+2}y_{j+2})^{\varepsilon}, b(x_{i+2}u_{k+2})^{\varepsilon}v_{l+1}^{\nu}, F(i,j,k,l,1,0,1)\rangle, \quad j \leq k; \\
\langle a, bv_{l+1}^{\nu}, F(i,j,k,l,1,0,-1)\rangle, \quad k \leq l;
\end{array}$$

for $i,j \gtrsim 0,\ k,l \gtrsim 0$ *and* $\alpha \in \{-1,1\}$:

$$\begin{array}{l}
\langle ax_{i+1}^{\varepsilon}, b, F(i,j,k,l,1,1)\rangle, \quad k \le l;\\
\langle a(x_{i+2}y_{j+2})^{\varepsilon}, b(x_{i+2}u_{k+2})^{\gamma} v_{l+2}^{\gamma-\varepsilon}, F(i,j,k,l,1,1,1)\rangle, \quad j < k < l;\\
\langle a(x_{i+2}y_{j+2})^{\varepsilon\gamma}, b(x_{i+2}u_{k+2})^{\gamma(1-\varepsilon)} v_{k+2}^{\gamma}, F(i,j,k,k,1,1,1)\rangle, \quad j \ne k;\\
\langle a, b(x_{i+2}u_{j+2}v_{j+2})^{\gamma}, F(i,j,j,j,1,1,1)\rangle;\\
\langle a, b, F(i,j,k,l,1,1,2,\alpha)\rangle, \quad k \le l, \ \textit{and}\ j \le k\ \textit{if}\ i \ne 0.
\end{array}$$

Proof. Observe first that we have here a sublist of the list given in Theorem 3.2.9, and also (by Theorem 1.3.5 and Lemma 3.1.9) this sublist consists of irreducible groups. The rest of the proof is essentially comparison with the list in Theorem 3.2.9, once we know how to recognise intersections of the BS_4-conjugacy classes with the set of irreducible groups G in that list. That is, we need to recognise when G^{sz} is also in the list, for some $s \in S_3 = \langle (1,2),(2,3)\rangle$ and $z \in BV_4$. Since $B \cap G^s \trianglelefteq BG^s = (BG)^s = (BV_4)^s = BV_4$, we have $B \cap G^{sz} = (B \cap G^s)^z = B \cap G^s$. Under the assumption that G and G^{sz} are listed in Theorem 3.2.9, this implies that $B \cap G$ and $B \cap G^s = (B \cap G)^s$ are in the modified list of Proposition 3.1.12. From the orbit list given in the proof of Proposition 3.1.12 we see that, for $B \cap G$ in the first seven lines, this restricts s to two choices, but otherwise all six choices are allowed. For each non-trivial choice of s so allowed, we have to calculate G^s and then find the unique BV_4-conjugate (in fact B-conjugate) of G^s in the list of Theorem 3.2.9. This identifies all the BS_4-conjugates of G in that list, and then inspection would show that precisely one of them is listed in the statement of the present theorem.

We illustrate the described procedure with two particular examples, and omit the similar calculations for all other choices of G. First, let

$$G = \langle ax_{i+1}^{\varepsilon} y_{j+1}^{\eta}, bx_{i+1}^{\gamma}, F(i,j,k,l,0,1)\rangle.$$

From the orbit list in the proof of Proposition 3.1.12, we see that $s \in \langle (1,2)\rangle$. Now

$$\begin{aligned}
G^{(1,2)} &= \langle ax_{i+1}^{\varepsilon} y_{j+1}^{\eta}, abx_{i+1}^{\gamma}, F(i,j,l,k,0,1)\rangle\\
&= \langle ax_{i+1}^{\varepsilon} y_{j+1}^{\eta}, bx_{i+1}^{\gamma+\varepsilon} y_{j+1}^{\eta}, F(i,j,l,k,0,1)\rangle.
\end{aligned}$$

The superscript $\gamma+\varepsilon$ may be read modulo 2, since $x_{i+1}^2 \in F(i,j,k,l,0,1)$. The unique BV_4-conjugate of $G^{(1,2)}$ must be in the same line of the list of Theorem 3.2.9 as G. We have

$$G^{(1,2)} \sim_{y_{j+2}^{-\eta}} \langle ax_{i+1}^{\varepsilon} y_{j+1}^{\eta}, bx_{i+1}^{\gamma+\varepsilon}, F(i,j,l,k,0,1)\rangle,$$

which (with the informal convention that $\gamma+\varepsilon$ is read modulo 2) is visibly in the list of Theorem 3.2.9, as required. This group and G form the intersection of that list with one BS_4-conjugacy class, so we have to check that one and only one of them appears in the list above. This is clearly true when $k < l$ or $l < k$, with G appearing in the first case and $G^{(1,2)}$ in the second. When $k = l$ and $\varepsilon = 1$, the two groups differ only in their second stated generator. In one or the other group (depending on the value of γ) the generator is b; that group is there and the other is not. When $k = l$ and $\varepsilon = 0$, the two groups coincide and this one group is in the list.

The second example we consider is

$$G = \langle ax_{i+1}^{\varepsilon}, bx_{i+1}^{\gamma}u_{k+1}^{\mu}, F(i,j,k,l,0,0,0,-1)\rangle.$$

By inspection of the orbit list in the proof of Proposition 3.1.12, we see that in the first instance s can be any element of S_3. When j,k,l are pairwise distinct, the six S_3-conjugates of G differ even in their diagonal subgroups, but only by permutations of the j,k,l. Just one S_3-conjugate has those parameters in ascending order. Passing to the unique B-conjugates in the list of Theorem 3.2.9 has no effect on those three parameters. Hence, there we find six distinct groups, all in the third line, and just one has j,k,l in ascending order. It is not necessary to calculate how the values of ε,γ,μ for the six groups depend on the parameter values in the label of $B \cap G$, since it is evident from what has been said so far that precisely one of the groups is listed above.

The determination is slightly more complicated when two of the j,k,l are equal but distinct from the third. In this case, G has either three or six distinct conjugates in the list of Theorem 3.2.9, and among these only one or two have the three critical parameters in the order stipulated above. So we may assume without loss of generality that $j \neq k = l$; then the only relevant non-trivial conjugate to consider is $G^{(1,2)}$. We have

$$\begin{aligned}
G^{(1,2)} &= \langle ax_{i+1}^{\varepsilon}, abx_{i+1}^{\gamma}v_{k+1}^{\mu}, F(i,j,k,k,0,0,0,-1)\rangle \\
&= \langle ax_{i+1}^{\varepsilon}, bx_{i+1}^{\gamma+\varepsilon}v_{k+1}^{\mu}, F(i,j,k,k,0,0,0,-1)\rangle \\
&= \langle ax_{i+1}^{\varepsilon}, bx_{i+1}^{\gamma+\varepsilon}y_{j+1}^{\mu}u_{k+1}^{\mu}, F(i,j,k,k,0,0,0,-1)\rangle,
\end{aligned}$$

using the fact that $y_{j+1}u_{k+1}v_{k+1}^{-1} \in F(i,j,k,k,0,0,0,-1)$. Further conjugation by $y_{j+2}^{-\mu}$ produces

$$\langle ax_{i+1}^{\varepsilon}, bx_{i+1}^{\gamma+\varepsilon}u_{k+1}^{\mu}, F(i,j,k,k,0,0,0,-1)\rangle,$$

which is the unique B-conjugate of $G^{(1,2)}$ in the list of Theorem 3.2.9. We have to verify that precisely one of G and this group is listed above. When $\varepsilon = 1$, the two groups differ only in their second stated generator. In one or the other group (depending on the value of γ) the generator is bu_{k+1}^{μ}; that group is there and the other is not. When $\varepsilon = 0$, the two groups are the same and this one group is in the list.

When $j = k = l$, we need to determine, for each $s \in S_3$, the B-conjugate of G^s in the list of Theorem 3.2.9. These are already known for $s \in \langle(1,2)\rangle$:

$$\langle ax_{i+1}^{\varepsilon}, bx_{i+1}^{\gamma}u_{j+1}^{\mu}, F(i,j,j,j,0,0,0,-1)\rangle, \tag{3.18}$$

$$\langle ax_{i+1}^{\varepsilon}, bx_{i+1}^{\gamma+\varepsilon}u_{j+1}^{\mu}, F(i,j,j,j,0,0,0,-1)\rangle. \tag{3.19}$$

Next,

$$G^{(2,3)} = \langle ax_{i+1}^{\gamma} y_{j+1}^{\mu}, bx_{i+1}^{\varepsilon}, F(i,j,j,j,0,0,0,-1)\rangle,$$

which is conjugate by $(y_{j+2}u_{j+2}v_{j+2})^{\mu}$ to

$$\langle ax_{i+1}^{\gamma}, bx_{i+1}^{\varepsilon} u_{j+1}^{\mu}, F(i,j,j,j,0,0,0,-1)\rangle, \tag{3.20}$$

using the fact that $y_{j+1}u_{j+1}v_{j+1} \in F(i,j,j,j,0,0,0,-1)$. Similarly, after conjugating each of the groups (3.18)–(3.20) by $(1,3)$ and appropriate elements of B, we obtain the other three BS_4-conjugates of G in the list of Theorem 3.2.9, viz.

$$\begin{gathered}\langle ax_{i+1}^{\gamma+\varepsilon}, bx_{i+1}^{\gamma} u_{j+1}^{\mu}, F(i,j,j,j,0,0,0,-1)\rangle,\\ \langle ax_{i+1}^{\gamma}, bx_{i+1}^{\gamma+\varepsilon} u_{j+1}^{\mu}, F(i,j,j,j,0,0,0,-1)\rangle,\\ \langle ax_{i+1}^{\gamma+\varepsilon}, bx_{i+1}^{\varepsilon} u_{j+1}^{\mu}, F(i,j,j,j,0,0,0,-1)\rangle.\end{gathered}$$

For fixed μ and $(\varepsilon,\gamma) = (0,0)$, the six BS_4-conjugates of G are the same, and that group is in the list. For fixed μ and $(\varepsilon,\gamma) \neq (0,0)$, the BS_4-conjugacy class of G contains the three groups defined by taking $(\varepsilon,\gamma) = (1,0), (0,1)$ and $(1,1)$ in (3.18). Of these, only the one corresponding to $(\varepsilon,\gamma) = (1,0)$ is listed. □

Remark 3.3.2 With reference to the orbit list given in the proof of Proposition 3.1.12, it may be seen that distinct V_4-submodules of B appearing as diagonal subgroups of groups in the list of Theorem 3.3.1 are not S_4-conjugate.

Remark 3.3.3 By Remark 1.3.8, distinct groups in the list of Theorem 3.3.1 are not even $M(4)$-conjugate.

The next two lemmas will assist us in identifying groups in the list of Theorem 3.3.1 with a unique SCN4 subgroup.

Lemma 3.3.4 *Let G be a finite irreducible subgroup of BV_4, and set $B \cap G = B_1$. Suppose G has an* SCN4 *subgroup $S \neq B_1$, and set $Q = B_1 \cap S$. Then either*

(i) *$B_1 S \neq G$, in which case Q is a V_4-submodule of index 2 in B_1 such that $|\mathsf{C}_{V_4}(Q)| = 2$ and $BS = B\mathsf{C}_{V_4}(Q)$, and $B_1 S$ has precisely 3 abelian maximal subgroups; or*

(ii) *$B_1 S = G$, in which case $Q = Z(G)$ is a cyclic subgroup of index 4 in B_1.*

Proof. Certainly (i) or (ii) occurs.

(i) In this case, $Q = B_1 \cap S = B \cap S$ is maximal in both B_1 and S, and

$$|BV_4 : BS| = |BG : BS| = |BG : B|/|BS : B| = |G : B \cap G|/|S : B \cap S| = 2.$$

Then $BS \leq B\mathsf{C}_{V_4}(Q) \leq BV_4$ forces $B\mathsf{C}_{V_4}(Q) = BV_4$ or $B\mathsf{C}_{V_4}(Q) = BS$. The first possibility is ruled out, since it implies that Q is V_4-trivial and then $B_1 = F_a Q$ is not faithful. Thus $BC_{V_4}(Q) = BS$ and $|C_{V_4}(Q)| = 2$. Since $Q \leq Z(B_1S)$, an abelian maximal subgroup of B_1S must contain Q. Otherwise, B_1S would be an abelian maximal subgroup of G, contradicting the irreducibility of G, by Lemma 1.3.1 (ii). Therefore, the abelian maximal subgroups of B_1S are in one-to-one correspondence with those of $B_1S/Q \cong C_2 \times C_2$.

(ii) If $B_1S = G$ then $Q = B_1 \cap S \leq Z(G)$. Hence

$$16 \leq |G : Z(G)| \leq |G : Q| = |B_1 : Q||S : Q| = 16.$$

Thus $Q = Z(G)$ is a V_4-trivial subgroup of B_1 of index 4. □

Remark 3.3.5 In Lemma 3.3.4 (ii), G, as the product of two abelian normal subgroups, has nilpotency class 2.

Lemma 3.3.6 *Let G be a finite irreducible subgroup of BV_4 and set $B_1 = B \cap G$. Then G has an* SCN4 *subgroup $S \neq B_1$ if and only if B_1 has a V_4-submodule Q of index* 2 *such that $|C_{V_4}(Q)| = 2$ and $\Phi(G) \leq Q$.*

Proof. Suppose that S exists. Since $G/B_1 \cong G/S \cong V_4$, we have $\Phi(G) \leq B_1 \cap S$. In case (i) of Lemma 3.3.4, the conclusion in one direction follows. In case (ii), G has class 2 and B_1 cannot be cyclic. For if B_1 were cyclic then G would be isomorphic to one of the groups referred to in the sentence before Theorem 2 of [3]; but these groups all have class greater than 2. Thus $B_1 \geq F_t$ for some t and $Q = (B_1 \cap S)F_t$ satisfies the stated criteria. Conversely, suppose that Q exists, and write $C_{V_4}(Q) = \langle t \rangle$. An element of $tB \cap G$ has t-trivial square, which is therefore contained in Q, since B_1 is faithful. The normal subgroup S generated by Q and an element of $tB \cap G$ is clearly abelian. It is also self-centralising, by Lemma 1.3.1, Proposition 1.3.6 and the irreducibility of G. Finally, $\Phi(G) \leq Q < S$ implies that G/S is elementary abelian of order 4. □

If G and H are finite irreducible subgroups of BV_4 and θ is an isomorphism of G onto H such that $(B \cap G)\theta \neq B \cap H$, then the hypotheses of Lemma 3.3.4 are satisfied for G and $S = (B \cap H)\theta^{-1}$. In case (i) of the lemma we say that θ is a *type I isomorphism*, and in case (ii), a *type II isomorphism*. As we know, if two distinct groups in the list of Theorem 3.3.1 are conjugate, then each has more than one SCN4 subgroup, and the linear isomorphism between them is type I or type II. We treat the type I linear isomorphisms first. After some preliminary results, we describe in Proposition 3.3.9 the form of an element of $GL(4)$ that induces a type I linear isomorphism.

Lemma 3.3.7 *Suppose that G and H are finite irreducible subgroups of BV_4 and $\theta: G \to H$ is a type I isomorphism such that $C_{V_4}(Q) = C_{V_4}(Q\theta)$, where $Q = (B \cap G) \cap (B \cap H)\theta^{-1}$. Then $Q\theta = Q$.*

Proof. By hypothesis and Lemma 3.3.4 (i), both Q and $Q\theta$ are centralised by a single involution $t \in V_4$. Then it is not difficult to see that an element of $sB \cap G$ is mapped to an element of $s\langle t\rangle B \cap H$ under θ, where $s \neq t$ is an involution of V_4. Thus θ restricted to Q is a T-isomorphism, implying the result by Proposition 1.1.2. □

Lemma 3.3.8 *Let $Y_1 > B_0$ be a subgroup of Y. Then $\mathsf{N}_{GL(4)}(Y_1) = \mathsf{N}_{GL(4)}(Y)$ is the subgroup of $GL(4)$ generated by b and all block diagonal matrices of the form*

$$f = \begin{pmatrix} f_1 & 0 \\ 0 & f_2 \end{pmatrix}; \tag{3.21}$$

$f_1, f_2 \in GL(2)$.

Proof. Matrix multiplication. □

Proposition 3.3.9 *Suppose G and H are distinct groups in the list of Theorem 3.3.1 such that $e \in GL(4)$ induces a type I linear isomorphism of G onto H. By Lemma 3.3.4 (i), let s and t be the involutions of V_4 such that*

$$\mathsf{C}_{V_4}((B\cap G)\cap(B\cap H)^{e^{-1}}) = \langle t\rangle \quad \textit{and} \quad \mathsf{C}_{V_4}((B\cap G)^e\cap(B\cap H)) = \langle s\rangle.$$

Then either $e = rwfm$ or $e = rwbfm$, for some $r \in \langle(1,2),(2,3)\rangle, w \in UV$ and $m \in M(4)$, where f is of the form (3.21) with

$$f_1 = f_2 = \begin{pmatrix} 1 & -1 \\ 1 & 1 \end{pmatrix}.$$

Proof. We choose $r, \tilde{r} \in S_3$ such that $t^r = a = s^{\tilde{r}}$. Then $r^{-1}e\tilde{r}$ induces a type I linear isomorphism of G^r onto $H^{\tilde{r}}$. Setting

$$Q = (B\cap G^r)\cap(B\cap H^{\tilde{r}})^{\tilde{r}^{-1}e^{-1}r},$$

we have

$$\mathsf{C}_{V_4}(Q) = \mathsf{C}_{V_4}(Q^{r^{-1}e\tilde{r}}) = \langle a\rangle.$$

Hence $Q \leq XY$, and $Q^{r^{-1}e\tilde{r}} = Q$ by Lemma 3.3.7. We assert that $r^{-1}e\tilde{r} \in \mathsf{N}_{GL(4)}(Y)$. To see that this is true, suppose first that Q is noncyclic. Then $r^{-1}e\tilde{r}$ normalises $\Omega_1(Q) = \Omega_1(XY) = F_a$ and so fixes or interchanges its two nonscalar elements x_1y_1 and $x_1y_1^{-1}$. In either case, $r^{-1}e\tilde{r} \in \mathsf{N}_{GL(4)}(\langle y_1\rangle) = \mathsf{N}_{GL(4)}(Y)$. Suppose next that Q is cyclic. Then QX/X contains the unique minimal subgroup $\langle y_1\rangle X/X$ of XY/X. Thus $y_1^{r^{-1}e\tilde{r}} = xy_1$ for some $x \in X$; in fact $x \in B_0$, and so $r^{-1}e\tilde{r}$ again normalises $\langle y_1\rangle$.

For some $z_1 \in XY$ and $z_2 \in UV$, we have $(B\cap H^{\tilde{r}})^{\tilde{r}^{-1}e^{-1}r} = \langle az_1z_2, Q\rangle$. Choose $w \in UV$ such that $w^2 = z_2$. Since $\langle az_1z_2, Q\rangle^{r^{-1}e\tilde{r}} \leq B$ and $r^{-1}e\tilde{r} \in \mathsf{N}_{GL(4)}(XY)$, we see that $a^{wr^{-1}e\tilde{r}} \in \Omega_1(B)$. Certainly $wr^{-1}e\tilde{r} \in \mathsf{N}_{GL(4)}(Y)$ and by Lemma 3.3.8 is consequently either f or bf, where f is of the form (3.21) (the group of all matrices of the form (3.21) in $GL(4)$ is normalised by b). Now $a^f = a^{bf}$ is a diagonal matrix with nonzero entries ± 1. Since a itself is of the form (3.21), with blocks $(1,2)$, it follows that $(1,2)^{f_i}$ is one of the diagonal matrices $(1,-1)$ or $(-1,1)$. Hence f_i is

$$\begin{pmatrix} -\alpha & \beta \\ \alpha & \beta \end{pmatrix} = \begin{pmatrix} 1 & -1 \\ 1 & 1 \end{pmatrix} (1,2) \begin{pmatrix} \alpha & 0 \\ 0 & \beta \end{pmatrix} \quad \text{or} \quad \begin{pmatrix} \alpha & -\beta \\ \alpha & \beta \end{pmatrix} = \begin{pmatrix} 1 & -1 \\ 1 & 1 \end{pmatrix} \begin{pmatrix} \alpha & 0 \\ 0 & \beta \end{pmatrix}$$

for some $\alpha, \beta \in \mathbb{C}^\times$. That is, the blocks of f are as stated, up to postmultiplication of f by a monomial matrix. □

In the next theorem, we determine all type I linear isomorphisms.

Theorem 3.3.10 *There is a type I linear isomorphism from a group in the list of Theorem 3.3.1 with diagonal subgroup $F(i,j,0,0,0,1,\xi)$, $i,j \geq 1$ and $\xi \in \{-1,0\}$, to a group in the list with diagonal subgroup $F(i,j,1,1,0,0)$ or $F(i,j,0,0,0,1,1)$. The only other type I linear isomorphisms between groups in the list are:*

$$\begin{aligned} &\langle a,b,F(i,j,0,0,0,1,1)\rangle \sim \langle ax_{i+1},b,F(i,j,1,1,0,0)\rangle; \\ &\langle a,b,F(i,j,0,0,1,1,1)\rangle \sim \langle ax_{i+1},b,F(i,j,1,1,1,0)\rangle, \ i,j \geq 1; \\ &\langle a,b,F(0,0,0,0,1,1,1)\rangle \sim \langle a,b,F(0,0,1,1,1,0)\rangle. \end{aligned}$$

Proof. Let G,H,e,t,s and f be as in the statement of Proposition 3.3.9. We repeat some of the reasoning in the proof of that proposition (with the same notation carried over), distinguishing between the cases $t=a$ and $t \neq a$. Preparatory to this, observe that

$$a^f = x_1 u_1^{-1}, \tag{3.22}$$
$$b^f = b, \tag{3.23}$$
$$u_1^f = a x_1^{-1}. \tag{3.24}$$

Also, if we write $G = \langle ax_a y, bx_b uv, B \cap G\rangle$ then the following conditions are equivalent to $\Phi(G) \leq Q^{r^{-1}}$:

$$x_a^2 y^2 \in Q^{r^{-1}}, \tag{3.25}$$
$$x_b^2 u^2 \in Q^{r^{-1}}, \tag{3.26}$$
$$y^{-2} u^2 v^2 \in Q^{r^{-1}}. \tag{3.27}$$

Case 1: $t=a$. Here, $r=1$ and $Q \leq XY$. Maximality of Q and faithfulness of $B \cap G$ then imply that $Q = G \cap XY$ and Q is noncyclic. Note that (3.25) is automatically satisfied in this case.

In Example 3.1.11 we determined the finite noncyclic V_4-submodules of B that are centralised by $\langle a\rangle$, and, with reference to Figures 3.1 and 3.2, the V_4-submodules which contain them maximally. By considering only the faithful ones among the latter, we see that either $Q = F(i,j,0,0,0,1)$ and $B\cap G$ is $F(i,j,1,1,0,0)$ or one of the $F(i,j,0,0,0,1,\xi)$; or $Q = F(i,j,0,0,1,1)$ and $B\cap G$ is one of $F(i,j,0,0,1,1,1)$ or $F(i,j,1,1,1,0)$. In all cases, $u_1 \in (B\cap G)XY$. If $g \in G\cap XYu_1$ then

$$\langle ax_a y, Q\rangle, \quad \langle ax_a yg, Q\rangle, \quad B\cap G$$

are three distinct abelian maximal subgroups of $\langle ax_a y, B\cap G\rangle = (B\cap G)(B\cap H)^{e^{-1}}$. By Lemma 3.3.4 (i), we have therefore produced explicitly the two possibilities for $(B\cap H)^{e^{-1}}$. Then we may choose $w \in UV$ as in the proof of Proposition 3.3.9 to be u_2^τ, for some $\tau \in \{0,1\}$. By the proposition, we conclude that $e = b^\sigma u_2^\tau fm$ for some $\sigma \in \{0,1\}$ and $m \in M(4)$.

All type I linear isomorphisms between groups in the list of Theorem 3.3.1 will be determined in the following way. First, we select G in the list with $B\cap G$ one of the options identified above and such that G has an SCN4 subgroup different from $B\cap G$: by Lemma 3.3.6, this means checking whether (3.26) and (3.27) are satisfied for the corresponding Q. Then we calculate the conjugates $G^{b^\sigma u_2^\tau f}$. (Of course these conjugates all contain F_a, and it is true but perhaps not obvious that they all lie in BS_4.) Finally, we identify the (unique) B-conjugate of each of these groups in the list of Theorem 3.3.1. Since distinct groups in the list of Theorem 3.3.1 are not $M(4)$-conjugate, this accounts for all type I linear isomorphisms. We proceed to apply the described method, assuming for the moment that $|Q| > 4$ and thus $\langle x_1, y_1\rangle \le Q$.

Suppose that $Q = F(i,j,0,0,0,1)$. When $B\cap G = F(i,j,0,0,0,1,\xi)$, $\xi \neq 1$, we verify that $\Phi(G) \le Q$ by (3.26), (3.27) and Theorem 3.3.1. Furthermore, using (3.22)–(3.24), we check that $G^f \le BS_4$ and $B\cap G^f$ is one of the following:

$$\begin{aligned}\langle ax_{i+1}^\varepsilon, x_i, y_j\rangle^f &= \langle x_{i+1}^\varepsilon u_1, x_i, y_j\rangle \\ &= \begin{cases} F(i,j,1,1,0,0) & \varepsilon = 0 \\ F(i,j,0,0,0,1,1) & \varepsilon = 1.\end{cases}\end{aligned}$$

By Remark 3.3.2, G^f is BS_4-conjugate to one of the groups in the list of Theorem 3.3.1 with diagonal subgroup $F(i,j,1,1,0,0)$ or $F(i,j,0,0,0,1,1)$. This proves the first claim of the theorem. It is not necessary to examine the other conjugates $G^{b^\sigma u_2^\tau f}$, since any type I isomorphisms arising from these will be determined at subsequent stages of this proof. When $B\cap G = F(i,j,0,0,0,1,1)$, the choices for G are restricted to the $\langle ay_{j+1}^\eta, b, B\cap G\rangle$, by (3.26) and Theorem 3.3.1. Since $G^b = G$ for these choices, we only need to calculate the two conjugates $G^{u_2^\tau f}$. If $\eta = 1$ then by (3.22)–(3.24),

$$\begin{aligned} G^{u_2^\tau f} &= \langle ay_{j+1}u_1^\tau, b, x_{i+1}u_1, x_i, y_j\rangle^f \\ &= \begin{cases} \langle ax_{i+1}, b, F(i,j,0,0,0,1,0)\rangle & \tau = 0 \\ \langle ax_{i+1}, b, F(i,j,0,0,0,1,-1)\rangle & \tau = 1; \end{cases} \end{aligned}$$

linear isomorphisms that have already been implicitly noted. If $\eta = 0$ then

$$G^{u_2^\tau f} \sim_B \begin{cases} \langle ax_{i+1}, b, F(i,j,1,1,0,0)\rangle & \tau = 0 \\ G & \tau = 1. \end{cases}$$

The first of these linear isomorphisms is given in the statement of the theorem. When $B \cap G = F(i,j,1,1,0,0)$, $G = G^b$ must be one of the $\langle ax_{i+1}^\varepsilon y_{j+1}^\eta, bx_{i+1}^{\gamma(1-\varepsilon)}, B \cap G\rangle$, by (3.26), (3.27) and Theorem 3.3.1. Then

$$G^{u_2^\tau f} = \langle a, bx_{i+1}^{\gamma(1-\varepsilon)}, x_{i+1}^\varepsilon y_{j+1}^\eta u_1, x_i, y_j\rangle.$$

In fact, $G^{u_2^\tau f} = G$ (this happens when $\varepsilon = \eta = 0$), or $u_2^\tau f$ induces a linear isomorphism already noted.

Now suppose that $Q = F(i,j,0,0,1,1)$. When $B \cap G = F(i,j,0,0,1,1,1)$, it is clear by (3.26), (3.27) and Theorem 3.3.1 that G must be the split extension. We have $G^b = G^{u_2 f u_2} = G$, so that only G^f need be considered:

$$G^f = \langle ax_{i+1}, b, F(i,j,1,1,1,0)\rangle.$$

This is the second of the conjugacies explicitly given in the statement of the theorem. When $B \cap G = F(i,j,1,1,1,0)$, our method yields no new isomorphisms.

To complete the analysis of this case, let $Q = F_a$. By Example 3.1.11 and the usual reasoning, $G = G^b$ must be either $\langle a, b, F(0,0,0,0,1,1,1)\rangle$ or one of the two groups $\langle ax_1^\varepsilon, b, F(0,0,1,1,1,0)\rangle$. Conjugation by $u_2^\tau f$ yields the last of the explicitly listed linear isomorphisms, and no new ones.

Case 2: $t \neq a$ (and $s \neq a$, by symmetry). In this case, $Q = (B \cap G^r) \cap (B \cap H)^{e^{-1}r}$ for $r = (2,3)$ or $r = (1,3)$. In fact, Q may be assumed cyclic. To see this, suppose that Q is noncyclic. The possible choices for Q and then $B \cap G^r$ were given in Case 1. There we saw that each faithful finite V_4-submodule of B containing Q maximally occurs as the diagonal subgroup of a group listed in Theorem 3.3.1. Hence $(B \cap G)^r = B \cap G$ by Remark 3.3.2. Inspection of the orbit list in the proof of Proposition 3.1.12 shows that only $F(i,1,1,1,0,0)$ and $F(0,0,0,0,1,1,1)$ from the collection of possibilities for $B \cap G$ are normalised by r. By Remark 3.3.11 below, the latter possibility may be disregarded. Otherwise, for $r = (2,3)$ and $r = (1,3)$ in turn, we apply the same method as in Case 1 (note that w may still be chosen as u_2^τ) for each G in the list of

Theorem 3.3.1 satisfying (3.25)–(3.27) and with $F(i,1,1,1,0,0)$ as diagonal subgroup. No new isomorphisms are obtained.

Henceforth, Q is cyclic. Then since Q is maximal in both $(B\cap G)^r$ and $(B\cap H)^{e^{-1}r}$, and $Q^{r^{-1}e\tilde{r}}=Q$, it follows that $B\cap G=F_aQ^{r^{-1}}$ and $B\cap H=F_aQ^{\tilde{r}^{-1}}$. These two V_4-submodules of B are conjugate by $(1,2)$, and so $B\cap G=B\cap H$ by Remark 3.3.2. To construct the possibilities for $B\cap G$, we determine the finite cyclic V_4-submodules M of B such that $\mathsf{C}_{V_4}(M)=\langle ab\rangle$, say, and then for each such M identify which one of F_aM or $F_aM^{(1,2)}$ appears as the diagonal subgroup of a group in the list of Theorem 3.3.1. Reasoning as in the proof of Lemma 3.1.1, we see that the possibilities for M are

$$\langle v_l\rangle,\ \ \langle x_1v_{l+1}\rangle,\ \ \langle x_{l+1}v_1\rangle;\ \ l\geq 1.$$

Accordingly, $B\cap G$ is one of $F(0,0,1,l,1,0)$, $F(0,0,l,1,1,0,1)$ or $F(l,1,0,0,0,1,1)$. In the second or third situation, the only group listed in Theorem 3.3.1 that satisfies (3.25)–(3.27) is the split extension of $B\cap G$ by V_4, and so there is nothing to prove. In the first situation, G must be $G_{\varepsilon,\nu}=\langle ax_1^{\varepsilon},bv_{l+1}^{\nu},B\cap G\rangle$, for some $\varepsilon,\nu\in\{0,1\}$. A short calculation reveals that the Frattini quotient of $G_{\varepsilon,0}$ has rank 4, and that of $G_{\varepsilon,1}$ has rank 3. Also, $G_{1,1}$ and $G_{0,1}$ are isomorphic to $Q^+(2^{l+4})$ and $D^+(2^{l+4})$ respectively, as defined in [3]. It is shown in [3] that $Q^+(2^{l+4})\not\cong D^+(2^{l+4})$. A group with $\nu=0$ has Frattini quotient of rank 4, whereas a group with $\nu=1$ has Frattini quotient of rank 3. So for this choice of $B\cap G$, the only groups that are possibly isomorphic are $G_{0,0}$ and $G_{1,0}$. But $G_{0,0}$ splits over each of its three SCN4 subgroups, whereas $G_{1,0}$ does not split over $B\cap G$. This completes our examination of Case 2 and hence the proof of the theorem. □

Remark 3.3.11 A group of order 32 with an SCN4 subgroup and centre of order 2 has a faithful irreducible representation of degree 4. There are seven isomorphism classes of such groups (see [11]). Since there are eight groups in the list of Theorem 3.3.1 of order 32, the last line in the statement of Theorem 3.3.10 shows that we have accounted for all isomorphisms between such groups.

All that remains now in our solution of the conjugacy problem is to deal with the type II linear isomorphisms. In fact, the effect of these isomorphisms has already been determined, as we will show in Proposition 3.3.13 below.

Lemma 3.3.12 *Let G be a finite irreducible subgroup of BV_4 of class 2. Then*

(i) $[B\cap G,G]=B_0$.

If also G splits over $B\cap G$ then

(ii) $\exp(G)=\max\{\exp(B\cap G),4\}$.

Proof. (i) Since $[B \cap G, G] \leq [G, G] \leq Z(G) = G \cap X$ and $G^2 \leq B \cap G$, we have $1 = [z, g^2] = [z, g][z, g]^g = [z, g]^2$ for all $z \in B \cap G$ and $g \in G$. Hence $[B \cap G, G]$ lies in $\Omega_1(X) = B_0$. Faithfulness of $B \cap G$ then forces $[B \cap G, G] = B_0$.

(ii) Here, we may assume that $V_4 < G$, by Lemma 3.2.8. Now $(tz)^n = t^n z^n [z, t]^{\binom{n}{2}}$ for all $t \in V_4$, $z \in B \cap G$ and $n \geq 1$. Together with (i), this implies that if $|z| \geq 4$ then $(tz)^{|z|} = 1$. On the other hand, if $\exp(B \cap G) = 2$ then $|tz| \leq 4$ for all t and z. But G is not abelian and so has at least one element of order 4, which is therefore the exponent of G in this case. □

Proposition 3.3.13 *Modify the list of Theorem* 3.3.1 *by omitting each group with diagonal subgroup one of the* $F(i, j, 0, 0, 0, 1, -1)$ *or the* $F(i, j, 0, 0, 0, 1, 0)$. *If there is a type II linear isomorphism between two groups in this modified list, then there is a type I linear isomorphism between them.*

Proof. Let G be a group in the modified list involved in a type II linear isomorphism. Since G has class 2,

$$[g_1 g_2, g_3 g_4] = [g_1, g_3]\,[g_1, g_4]\,[g_2, g_3]\,[g_2, g_4]$$

for all $g_i \in G$, $1 \leq i \leq 4$. From this and Lemma 3.3.12 (i), it follows that $[az, bw] \in B_0$ for all $az \in aB \cap G$ and $bw \in bB \cap G$. Thus, if we write G as $\langle a x_a y, b x_b u v, B \cap G \rangle$, then $[ay, buv] \in B_0$ and so

$$y^2, u^2, v^2 \in B_0. \tag{3.28}$$

By Lemma 3.3.4 (ii), $Z(G) = \langle x_i \rangle$ has index 4 in $B \cap G$, and by Remark 3.3.11 we may assume that $i \geq 1$. Hence $Z(G)F_a = F(i, 1, 0, 0, 0, 1)$ is a maximal subgroup of $B \cap G$ centralised by $\langle a \rangle$. By Example 3.1.11 and given the proposed omissions, $B \cap G$ is either $F(i, 1, 1, 1, 0, 0)$ or $F(i, 1, 0, 0, 0, 1, 1)$. Referring to the list in Theorem 3.3.1, we realise that G must be one of the groups

$$\langle a, b, F(i, 1, 1, 1, 0, 0) \rangle, \quad \langle a x_{i+1}, b, F(i, 1, 1, 1, 0, 0) \rangle, \quad \langle a, b, F(i, 1, 0, 0, 0, 1, 1) \rangle$$

in order for (3.28) to be satisfied. By Theorem 3.3.10, there is a type I linear isomorphism between the second and third groups. However, the first and second groups are not isomorphic: by Lemma 3.3.12 (ii), the first has exponent 2^{i+1}, but the second contains the element $a x_{i+1}$ of order 2^{i+2}. □

Combining Proposition 3.1.5, Theorem 3.3.1, Theorem 3.3.10 and Proposition 3.3.13, we obtain the following solution of the listing problem in the case $T = V_4$.

Theorem 3.3.14 *Modify the list of Theorem* 3.3.1 *by omitting*

the $\langle a, b, F(i, j, 0, 0, 1, 1, 1) \rangle$ *for all* $i, j \gtrsim 0$, *the* $\langle a, b, F(i,j,0,0,0,1,1)\rangle$ *for all* $i,j \geq 1$,

each group with diagonal subgroup one of the $F(i,j,0,0,0,1,\xi)$, *where* $\xi \in \{-1,0\}$ *and* $i,j \geq 1$.

Then the distinct groups in this modified list constitute a complete and irredundant list of $GL(4)$*-conjugacy class representatives of the finite irreducible subgroups of* BV_4.

If two groups in the list of Theorem 3.3.14 have different diagonal subgroups, yet each has a unique SCN4 subgroup, then they are not isomorphic. In Chapter 6 we will show that this statement is still true when the requirement of different diagonal subgroups is removed.

Chapter 4

The case $T = C$

The development of results in this chapter is modelled closely on the development in Chapter 3. Some results, particularly those relating to the extension and conjugacy problems, are markedly easier to obtain in the case $T = C$ than in the case $T = V_4$. For instance, cohomology of the finite cyclic group C may be calculated without recourse to spectral sequences, employing the standard expression of such cohomology as sections of the relevant coefficient module. In the conjugacy problem for $T = C$, to find BS_4-conjugacy classes we only have to consider action on BT-conjugacy classes by $\langle a \rangle$, since $\mathsf{N}_{S_4}(C) = D$ and $BD = BC \rtimes \langle a \rangle$; in the problem for $T = V_4$, we had to contend with action on BT-conjugacy classes by the larger group S_3.

Much of the notation of Chapter 3 will be preserved. As we did for V_4-submodules, we will give the finite C-submodules of B by group generating sets in the x_k, y_k, u_k, v_k. Note that

$$y_k^c = v_k, \quad u_k^c = u_k^{-1}, \quad v_k^c = y_k^{-1}$$

for all $k \geq 0$.

For an arbitrarily large positive integer n, let $A = (\mathbb{Z}/2^n\mathbb{Z})C$. There is a unique noncyclic submodule of order 4 in A_A. This will be denoted by the same notation F_b used for the matching unique noncyclic C-submodule $\langle x_1 u_1, y_1 v_1 \rangle$ of order 4 in B.

Proposition 4.1 *Define* $F_b^+ = \langle 1 + c^2, c + c^3 \rangle / F_b$ *and* $F_b^- = \langle 1 - c^2, c - c^3 \rangle / F_b$. *Then*

(i) $F_b^\perp / F_b = F_b^+ \oplus F_b^-$;

(ii) $\mathcal{L}(F_b^+)$ *is isomorphic to the submodule lattice of the regular* $(\mathbb{Z}/2^{n-1}\mathbb{Z})C_2$*-module;*

(iii) F_b^- *is a uniserial* A*-module.*

Proof. Note that $b = c^2$ acts trivially on F_b^+ and invertingly on F_b^- (so that we can define a $C/\langle b \rangle$-action on F_b^+ but not on F_b^-). For (i) and (ii), see the proof of

Lemma 3.1.2. For (iii), see the proof of Lemma 1.2.1. We state the A-submodules of F_b^- for later reference: these are

$$\begin{aligned}\mathrm{rad}^{2i} F_b^- &= \langle 2^i(1-c^2), 2^i(c-c^3)\rangle / F_b,\\ \mathrm{rad}^{2i-1} F_b^- &= \langle 2^{i-1}(1-c-c^2+c^3), 2^{i-1}(1+c-c^2-c^3)\rangle / F_b,\end{aligned}$$

for $1 \leq i \leq n$. □

Proposition 4.2 *Let G be a finite irreducible subgroup of BC. Then the following are equivalent:*

(i) *G has a self-centralising normal subgroup of index 4, different from $B \cap G$;*

(ii) *$B \cap G$ has a C-submodule Q of index 2 such that $\mathsf{C}_C(Q) = \langle b \rangle$;*

(iii) *G is conjugate to a subgroup of BV_4;*

(iv) *G is isomorphic to a subgroup of BV_4.*

Proof. It is straightforward to construct explicitly the (equivalence classes of) faithful representations of C in $\mathrm{Aut}(C_{2^i})$, $i \geq 3$. These may then be used to verify that each faithful finite cyclic C-submodule of B has a $\langle b \rangle$-trivial maximal submodule. Therefore, when $B \cap G$ is cyclic, (ii) is true regardless of the truth or falsity of (i), (iii) and (iv). If $B \cap G$ is noncyclic then (i) $\Rightarrow$ (ii) by reasoning similar to that used in the proofs of Lemmas 3.3.4 and 3.3.6.

Now assume (ii) and choose $g \in cB \cap G$; then $H = \langle g^2, Q \rangle$ is a normal self-centralising subgroup of G. Since $g^4 \in G \cap B^{1+c+c^2+c^3} \leq G \cap X = Z(G) < Q$, we see that H has index 4 in G. In fact, $G/H \cong V_4$, and then (iii) is a consequence of Proposition 1.3.6.

The implications (iii) $\Rightarrow$ (iv) and (iv) $\Rightarrow$ (i) are transparent. □

We state separately a result deriving from the proof of Proposition 4.2.

Lemma 4.3 *Let G be a finite irreducible subgroup of BC and suppose that $B \cap G$ is cyclic. Then G is conjugate to a subgroup of BV_4.*

The direct decomposition

$$B/F_b = (XU/F_b) \times (YV/F_b) \tag{4.1}$$

of the C-module B/F_b is an analogue of the direct decomposition of $F_b^{\perp}/F_b$ given in Proposition 4.1 (i); XU/F_b and YV/F_b match F_b^+ and F_b^-, respectively. Using the information available to us after translation of Proposition 4.1 and its proof, Theorem 3.1 of [10] can be applied to (4.1) to determine the finite noncyclic C-submodules of B (cf. the argument in Section 3.1). The results of this process are summarised next.

Theorem 4.4 *The pairwise distinct submodules defined below are the finite noncyclic C-submodules of B:*

$$\begin{aligned}
C(i,j,k,\delta_1,\delta_2) &= \langle x_i, u_j, y_k, v_k, (x_{i+1}u_{j+1})^{\delta_1}, (y_{k+1}v_{k+1})^{\delta_2}\rangle,\\
C(i,j,k,0,0,\xi) &= \langle x_i, u_j, y_k, v_k, x_{i+1}^{\xi}u_{j+1}^{\xi'}y_{k+1}v_{k+1}\rangle,\\
C(i,j,k,0,1,\xi) &= \langle x_i, u_j, y_{k+1}v_{k+1}, v_k, x_{i+1}^{\xi}u_{j+1}^{\xi'}y_{k+1}\rangle,\\
C(i,j,k,1,0,1) &= \langle x_{i+1}u_{j+1}, u_j, y_k, v_k, x_{i+1}y_{k+1}v_{k+1}\rangle,\\
C(i,j,k,1,1,1) &= \langle x_{i+1}u_{j+1}, u_j, y_{k+1}v_{k+1}, v_k, x_{i+1}y_{k+1}\rangle,\\
C(i,j,k,1,0,2,\alpha) &= \langle x_{i+1}u_{j+1}, u_j, u_{j+1}y_{k+1}v_{k+1}, x_{i+2}u_{j+2}^{\alpha}y_{k+1}\rangle,\\
C(i,j,k,1,1,2,\alpha) &= \langle x_{i+1}u_{j+1}, u_j, y_k, u_{j+1}y_{k+1}, x_{i+2}u_{j+2}^{\alpha}y_{k+2}v_{k+2}\rangle,
\end{aligned}$$

where $\delta_1, \delta_2 \in \{0,1\}$, $\alpha \in \{-1,1\}$, $\xi \in \{-1,0,1\}$ and ξ' denotes the representative in $\{-1,0,1\}$ of the congruence class modulo 3 of $1-\xi$. Each finite noncyclic C-submodule of B has been given a unique label, written generically as $C(i,j,k,\delta,\xi,\alpha)$, where δ stands for δ_1, δ_2; ξ stands for a void, or $\xi \in \{-1,0,1\}$, or $\xi = 2$; and α stands for a void, or $\alpha \in \{-1,1\}$. The parameters i,j,k range over the non-negative integers as follows:

$i,j \geq 1$ or $i,j \gtrsim 0$ according to whether δ_1 is 0 or 1, and
$k \geq 1$ or $k \geq 0$ according to whether δ_2 is 0 or 1,
using Notation 3.1.7.

The submodules $C(i,j,k,\delta,\xi,\alpha)$ with δ, ξ, α fixed form a family; there are 16 families of submodules in all. As before, it is possible to read useful properties of a submodule off its label (such as the order and interval of definition).

Of course, D-modules in the list of Theorem 3.1.8 (necessarily of additive rank at least 3) are also in the list of Theorem 4.4, albeit with different labels. The finite noncyclic D-submodules of B, labelled under the scheme introduced in this chapter, are specified in the next result.

Lemma 4.5 *Except for the $C(i,j,k,1,0,2,\alpha)$, each submodule in the list of Theorem 4.4 is a D-module.*

Proof. A D-submodule of B is precisely a C-submodule normalised by $\langle a\rangle$; equivalently, the C-isomorphism corresponding to the submodule by Theorem 3.1 of [10] is an $\langle a\rangle$-isomorphism. This is true for all such isomorphisms except possibly those between elementary abelian sections of XU/F_b and YV/F_b with composition length 2. All of these are $\langle a\rangle$-isomorphisms except the ones giving rise to the $C(i,j,k,1,0,2,\alpha)$: $C(i,j,k,1,0,2,1)^a = C(i,j,k,1,0,2,-1)$. □

In light of Proposition 4.2, we wish to determine the faithful submodules B_1 in the list of Theorem 4.4 with a submodule Q of index 2 such that $\mathsf{C}_C(Q) = \langle b\rangle$. It is clear that $Q \leq XU$ is noncyclic, so $Q = C(i,j,0,\delta_1,1)$ for some i,j and δ_1. Then by arguing as in Example 3.1.11 and the proof of Proposition 3.1.10, we see that B_1 is $C(0,0,1,1,0)$ or $C(0,0,0,1,1,1)$ when $|Q| = 4$, and one of the $C(i,j,1,0,0)$, $C(i,j,1,1,0)$, $C(i,j,0,0,1,\xi)$, or $C(i,j,0,1,1,1)$ when $|Q| > 4$. Figure 4.1 displays the Hasse diagram of the page $\mathcal{L}(C(i+1,j+1,1,0,0)/C(i,j,0,0,1))$, in which the edge B_1/Q lies when $|Q| > 4$. Vertices with underlined labels are the possibilities for Q in the page; the other labelled vertices are the possibilities for B_1. (When $|Q| = 4$, B_1/Q is not an edge in a Hasse diagram of this form.)

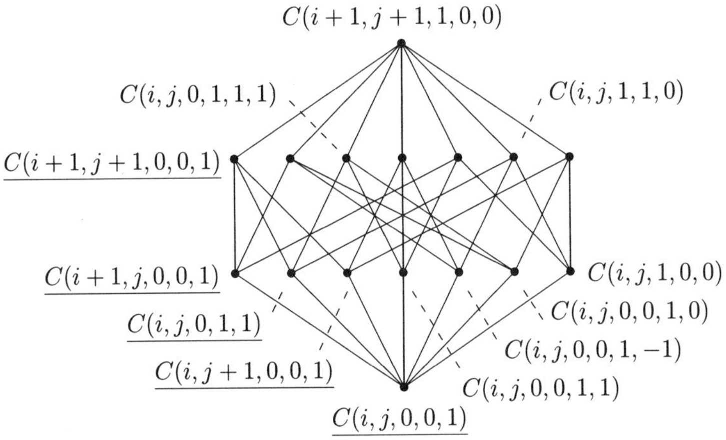

Figure 4.1: $\mathcal{H}(C(i+1,j+1,1,0,0)/C(i,j,0,0,1))$

The first part of the extension problem is solved by the following proposition.

Proposition 4.6 *Let B_1 be a finite noncyclic C-submodule of B. Then*

$$H^2(C,B_1) \cong \begin{cases} 1 & B_1 = C(i,j,k,1,0,2,\alpha) \text{ or } C(i,j,k,1,1,2,\alpha) \\ \mathbb{Z}_4 & B_1 = C(i,j,k,0,\delta_2) \text{ or } C(i,j,k,0,\delta_2,0) \\ \mathbb{Z}_2 & \text{otherwise.} \end{cases}$$

Proof. Denote by $\hat{c}$ the element $1+c+c^2+c^3$ of $\mathbb{Z}C$, and let ker refer to kernel on B_1. By Proposition 7.1, p.201 of [12], $H^2(C,B_1) \cong \ker(1-c)/B_1^{\hat{c}} = (B_1 \cap X)/B_1^{\hat{c}}$. Now c permutes cyclically the diagonal entries of an element of B, so that $\ker \hat{c} = B_1 \cap SL(4) = B_1 \cap YUV$. Since $B_1YUV = (B_1YUV \cap X)YUV$, it follows that $B_1^{\hat{c}} = (B_1YUV)^{\hat{c}} = (B_1YUV \cap X)^4$. The order of a generator of $B_1YUV \cap X$, and hence the order of the cyclic group $H^2(C,B_1)$, may be found by inspection of the generating sets given in Theorem 4.4. □

A proof of the next lemma is omitted; refer to the analogous result in the proof of Theorem 3.2.9.

Lemma 4.7 *Let G be a finite subgroup of BS_4 such that $BG = BC$. If x is any element of $c^{-1}GYUV \cap X$ then $x^4 \in Z(G)$ and $G \sim_B \langle cx, B \cap G\rangle$.*

We now present our solution of the listing problem in the case $T = C$.

Theorem 4.8 *As ε ranges over $\{0,1,2\}$, δ_2, η range over $\{0,1\}$, and α, ξ range over $\{-1,1\}$, the following is a list of finite irreducible subgroups of BC:*
for $i, j, k \geq 1$,

$$\begin{array}{ll} \langle cx_{i+2}^{\varepsilon}, C(i,j,k,0,\delta_2)\rangle, & k \geq 2 \text{ if } \delta_2 = 0; \\ \langle cx_{i+2}^{\varepsilon}, C(i,j,k,0,\delta_2,0)\rangle; & \\ \langle cx_{i+2}^{\eta}, C(i,j,k,0,\delta_2,\xi)\rangle; & \end{array}$$

for $i, j \gtrsim 0$ and $k \geq 1$,

$$\begin{array}{ll} \langle cx_{i+2}^{\eta}, C(i,j,k,1,\delta_2)\rangle, & k \geq 2 \text{ if } \delta_2 = 0; \\ \langle cx_{i+2}^{\eta}, C(i,j,k,1,\delta_2,1)\rangle; & \\ \langle c, C(i,j,k,1,0,2,1)\rangle; & \end{array}$$

for $i, j \gtrsim 0$ and $k \geq 0$,

$$\langle c, C(i,j,k,1,1,2,\alpha)\rangle.$$

If G is a finite irreducible subgroup of BC then either G is $GL(4)$-conjugate to a group in the list of Theorem 3.3.14, or BS_4-conjugate to a group in the preceding list. Furthermore, groups in the preceding list are pairwise non-isomorphic, and none of them is isomorphic to a group in the list of Theorem 3.3.14.

Proof. Set $B \cap G = B_1$. If B_1 is cyclic then G is conjugate to a group listed in Theorem 3.3.14, by Lemma 4.3. We assume for the rest of the proof that B_1 is noncyclic.

For some $i \geq 0$, $Z(G) = \langle x_i \rangle$. By Lemma 4.7, G is BC-conjugate to one of the groups $\langle cx_{i+2}^{n}, B_1\rangle$, $n \in \{0, \pm 1, 2\}$. If $|H^2(C, B_1)| = 1$ then by Lemma 3.2.8 we may take $n = 0$. If $|H^2(C, B_1)| = 2$ then $x_{i+1} \in B_1 YUV \cap X$; that is, $1 \in c^{-1}.cx_{i+1}B_1YUV$, and so $\langle cx_{i+1}, B_1\rangle \sim_B \langle c, B_1\rangle$ by Lemma 4.7. Similarly, $\langle cx_{i+2}^{-1}, B_1\rangle \sim_B \langle cx_{i+2}, B_1\rangle$ in this case. Then by Proposition 4.6, we see that G is BC-conjugate to one of the following groups:

$$\begin{array}{ll} \langle c, B_1\rangle & B_1 = C(i,j,k,1,0,2,\alpha) \text{ or } C(i,j,k,1,1,2,\alpha); \\ \langle cx_{i+2}^{\varepsilon}, B_1\rangle & B_1 = C(i,j,k,0,\delta_2) \text{ or } C(i,j,k,0,\delta_2,0); \\ \langle cx_{i+2}^{\eta}, B_1\rangle & \text{otherwise}; \end{array} \tag{4.2}$$

where ε and η range over $\{0,1,2,-1\}$ and $\{0,1\}$, respectively. By Theorem 2.5 and Proposition 4.6, these groups are pairwise non-conjugate in BC.

Next, we modify the list (4.2) by omitting reducible groups and (by Proposition 4.2) those with more than one self-centralising normal subgroup of index 4. This amounts to restricting parameter ranges so that the following C-submodules of B do not appear as diagonal subgroups: $C(0,0,0,1,1)$, $C(0,0,1,1,0)$, $C(0,0,0,1,1,1)$, and submodules corresponding to labelled vertices in Figure 4.1, for some $i,j \geq 1$. By Proposition 4.2, no group in this modified list is isomorphic to a subgroup of BV_4. The same is then true of the final list, since it is a sublist of the current one.

Now we select a single representative from each $\langle a \rangle$-conjugacy class of groups in the current list, thereby obtaining a list of groups that are pairwise non-conjugate in BS_4. By Lemma 4.5, we omit each group with one of the $C(i,j,k,1,0,2,-1)$ as diagonal subgroup. If $B_1 = C(i,j,k,0,\delta_2)$ or $C(i,j,k,0,\delta_2,0)$ then $\langle cx_{i+2}, B_1\rangle^a = \langle c^{-1}x_{i+2}, B_1\rangle = \langle cx_{i+2}^{-1}, B_1\rangle$. The range of ε in (4.2) is therefore restricted to $\{0,1,2\}$.

After carrying out all of the indicated modifications, we obtain the list in the statement of the theorem. Everything has been proved except the assertion that groups in this list are pairwise non-isomorphic. So let G and H be in the list, and suppose that $\theta: G \to H$ is an isomorphism. Then $B \cap G \sim_{BD} B \cap H$ by Theorem 2.7 (remember that groups with more than one self-centralising normal subgroup of index 4 are no longer present). But we know that $B \cap G$ is normalised by BD: thus, $B \cap G = B \cap H$ and the isomorphism question here is the restricted isomorphism question as posed in Chapter 2. It suffices to answer this question only when $H^2(C, B_1) \cong \mathbb{Z}_4$ for a finite noncyclic C-submodule B_1 of B. In that case, for each B_1, there are three groups in the list with diagonal subgroup B_1. Now the equivalence class of

$$1 \to B_1 \overset{\text{inc.}}{\to} \langle cx, B_1\rangle \overset{\text{proj.}}{\to} C \to 1$$

corresponds to $x^4 B_1^{\hat{c}} \in (B_1 \cap X)/B_1^{\hat{c}}$. Hence, of the three groups in the list of this theorem with diagonal subgroup B_1, one corresponds to a trivial 2-cocycle class, one to a class of order 2, and one to a class of order 4. These obviously lie in different orbits under the action of $\mathrm{Comp}(\chi) \leq \mathrm{Aut}(B_1) \times \mathrm{Aut}(C)$ on $H^2(C, B_1)$, and so we are done by Theorem 2.1. □

We would expect the listing problem for finite irreducible p-subgroups of $GL(p^n)$—that is, for finite irreducible p-subgroups of $M(p^n)$—always to be most tractable for groups with cyclic projection group. This is because of the lower degree of difficulty of the extension and conjugacy subproblems and the (restricted) isomorphism question. Certainly, the methods employed in this chapter may be applied with even greater facility to the case $n = 1$, providing an alternative to the approach taken in [5].

Chapter 5

The case $T = D$

As far as it is feasible to do so, in this chapter we will suppress reasoning along lines familiar from previous chapters.

Our first major result is a complete list of BS_4-conjugacy class representatives of the finite subgroups G of BD such that $BG = BD$. This list will then be refined to a list of $GL(4)$-conjugacy class representatives, excluding classes which have a representative in the list of Theorem 3.3.14 or Theorem 4.8.

If $B \cap G$ is a faithful D-module then G is irreducible, but the converse is not always true. For example, let B_1 be a noncyclic finite D-submodule of B upon which $\langle b \rangle$ acts trivially. Then $\langle ay_1, c, B_1 \rangle$ is irreducible, because its centraliser in $GL(4)$ is scalars. However, we may ignore non-faithful D-submodules by Proposition 5.2 below.

Lemma 5.1 *Suppose that G is a finite irreducible subgroup of BD such that $BG = BD$ and $B \cap G$ is not a faithful D-module. Then $\mathsf{C}_D(B \cap G) = \langle ac, b \rangle$.*

Proof. The normal subgroup $\mathsf{C}_D(B \cap G)$ of D must be one of $\langle b \rangle$ or $\langle ac, b \rangle$: if it were one of the other non-trivial normal subgroups of D, then $B \cap G$ would be scalars and G would have an abelian maximal subgroup. Now note that if $b = (1,3)(2,4)$ acts trivially on $B \cap G$ then so too does $ac = (1,3)$. □

Proposition 5.2 *In Lemma 5.1, G is conjugate to a subgroup of BV_4.*

Proof. Let π be canonical projection of G onto D. Choose $g \in b\pi^{-1}$: then $[g, G]\pi \leq [Z(D), D] = 1$. Thus $[g, G] \leq B \cap G$ and $H = \langle g, B \cap G \rangle$ is a normal abelian subgroup of G. Moreover, since π induces isomorphisms $G/B \cap G \cong D$ and $H/B \cap G \cong Z(D)$, we have $G/H \cong D/Z(D) \cong V_4$. The result follows from Proposition 1.3.6. □

Let B_1 be a faithful finite D-submodule of B. Note that B_1 is noncyclic and hence is in the list of Theorem 4.4. The $\langle b \rangle$-trivial D-submodules in this list are the $C(i, j, 0, \delta_1, 1)$. After deleting these and the $C(i, j, k, 1, 0, 2, \alpha)$, by Lemma 4.5, we obtain a list of all possibilities for B_1.

We proceed to calculate $|H^2(D,B_1)|$ by application of Propositions 3.2.1–3.2.3; in those results, e and f are replaced by c and a, respectively.

For $n \geq 1$, $H^n(C,B_1) \cong \ker \hat{c}/B_1^{1-c}$ (n odd) and $H^n(C,B_1) \cong \ker(1-c)/B_1^{\hat{c}}$ (n even), so that $|H^n(C,B_1)| = |H^{n+1}(C,B_1)|$. Then by Proposition 4.6 we have $|H^n(C,B_1)| = 1$ for $B_1 = C(i,j,k,1,1,2,\alpha)$. To find $|H^2(D,B_1)|$ in this case we use the following result, derived from a long exact sequence in the theory of LHS spectral sequences (see Exercise 3, p.355 of [15]): if K is a group, M a K-module and N a normal subgroup of K such that $H^n(N,M) = 0$, then $H^2(K,M) \cong H^2(K/N, M^N)$, where M^N is the set of N-fixed points in M. Thus

$$|H^2(D, C(i,j,k,1,1,2,\alpha))| = |H^2(\langle a\rangle, \langle x_i\rangle)| = 2.$$

Henceforth, we assume that $B_1 \neq C(i,j,k,1,1,2,\alpha)$ in our calculation of $|H^2(D,B_1)|$.

The domains of $d_2^{0,1}$ and $d_2^{1,1}$ are sections of $H^1(C,B_1) \cong (B_1 \cap YUV)/B_1^{1-c}$, whose structure is selectively displayed in Table 5.1. In the table, $[z]$ denotes the coset zB_1^{1-c}.

B_1	$H^1(C,B_1)$	
	isotype	generators
$C(i,j,k,0,0)$	$\mathbb{Z}_2 \times \mathbb{Z}_2$	$[u_j], [y_k]$
$C(i,j,k,0,1)$		$[u_j], [y_{k+1}v_{k+1}]$
$C(i,j,k,1,0)$	$\mathbb{Z}_2$	$[y_k]$
$C(i,j,k,1,1)$		$[y_{k+1}v_{k+1}]$
$C(i,j,k,0,0,1)$		$[u_j]$
$C(i,j,k,0,0,-1)$		
$C(i,j,k,0,1,1)$		
$C(i,j,k,0,1,-1)$		
$C(i,j,k,1,0,1)$		$[u_{j+1}y_{k+1}v_{k+1}]$
$C(i,j,k,1,1,1)$		$[u_{j+1}y_{k+1}]$
$C(i,j,k,0,1,0)$	$\mathbb{Z}_4$	$[u_{j+1}y_{k+1}]$
$C(i,j,k,0,0,0)$		$[u_{j+1}y_{k+1}v_{k+1}]$

Table 5.1: $H^1(C,B_1)$ as a section of B_1

Proposition 5.3 *For all $B_1 \neq C(i,j,k,1,0,1)$ in Table 5.1, $|\operatorname{coker} d_2^{0,1}| = 2$. If $B_1 = C(i,j,k,1,0,1)$ then $|\operatorname{coker} d_2^{0,1}| = 1$.*

Proof. By Proposition 3.2.1, $d_2^{0,1}$ maps into $(B_1 \cap X)/(B_1 \cap X)^2 \cong \mathbb{Z}_2$, and is defined by $zB_1^{1-c} \mapsto \overline{z}^{1+a}(B_1 \cap X)^2$, for $z \in B_1 \cap YUV$ and $\overline{z} \in B_1$ such that $z^{1+ac} =$

$\overline{z}^{1-c}$. Therefore, if B_1 has the property that $(B_1 \cap YUV)^{1+ac} \leq (B_1 \cap UV)^{1-c}$ then $d_2^{0,1} = 0$. Since $u_j^{1+ac} = u_j^{1-c}$ and $y_k^{1+ac} = (y_{k+1}v_{k+1})^{1+ac} = v_k^{1-c}$, with reference to Table 5.1 we see that B_1 has this property in all cases except $B_1 = C(i,j,k,0,0,0)$ and $B_1 = C(i,j,k,1,0,1)$. In the first case, $E_2^{0,1} = H^1(C,B_1)^2$ and $(u_{j+1}y_{k+1}v_{k+1})^{2+2ac} = (u_j v_{k-1})^{1-c} \in (B_1 \cap UV)^{1-c}$, so that in fact $d_2^{0,1} = 0$ here also. However, in the second case, $(u_{j+1}y_{k+1}v_{k+1})^{1+ac} = (x_{i+1}u_{j+1}v_k)^{1-c} \in (B_1 \cap XUV)^{1-c}$ and $(x_{i+1}u_{j+1}v_k)^{1+a}$ is a generator of $B_1 \cap X$, so that $d_2^{0,1}$ is surjective. This completes the proof. □

Proposition 5.4 *For all B_1 in Table 5.1, $\ker d_2^{1,1} = E_2^{1,1}$. Furthermore, if $B_1 = C(i,j,k,0,1,0)$ or $B_1 = C(i,j,k,0,0,0)$ then $|E_2^{1,1}| = 2$; in all other cases $E_2^{1,1} = H^1(C,B_1)$.*

Proof. Since $y_{k+1}v_{k+1}, u_j \in \ker(1-ac)$ and $y_k^{1-ac} = y_k^{1-c} = v_k^{ac-1}$, from Table 5.1 it is clear that $(B_1 \cap YUV)^{1-ac} \leq (B_1 \cap XY)^{1-c}$ for all $B_1 \neq C(i,j,k,0,1,0)$ (note that when $B_1 = C(i,j,k,1,1,1)$, we have $[u_{j+1}y_{k+1}] = [x_{i+1}y_{k+1}]$). For all such B_1, $\ker d_2^{1,1} = E_2^{1,1}$ by Proposition 3.2.2. If $B_1 = C(i,j,k,0,1,0)$ then $(B_1 \cap YUV)^{1-ac} \not\leq B_1^{1-c}$ but $(B_1 \cap YUV)^{2-2ac} \leq (B_1 \cap Y)^{1-c}$, so that $\ker d_2^{1,1} = E_2^{1,1} = H^1(C,B_1) \cong \mathbb{Z}_2$. Suppose that $B_1 \neq C(i,j,k,0,1,0)$ and $B_1 \neq C(i,j,k,0,0,0)$: by the proof of Proposition 5.3, the first sentence of this proof, and Proposition 3.2.2,

$$E_2^{1,1} = (B_1 \cap YUV)/(B_1 \cap YUV)^{1+ac}B_1^{1-c} = (B_1 \cap YUV)/B_1^{1-c} = H^1(C,B_1).$$

If $B_1 = C(i,j,k,0,0,0)$ then $E_2^{1,1} = (B_1 \cap YUV)/\langle u_j, y_k, v_k\rangle \cong \mathbb{Z}_2$. We are done. □

Proposition 5.5 *For all B_1 in Table 5.1, $\ker d_3^{0,2} = E_3^{0,2}$. Furthermore, $|E_3^{0,2}| = 2$ except when $B_1 = C(i,j,k,0,1,-1)$, in which case $|E_3^{0,2}| = 1$.*

Proof. The domain $E_3^{0,2}$ of $d_3^{0,2}$ is a subgroup (namely $\ker d_2^{0,2}$) of $H^2(C,B_1)$, which is known by Proposition 4.6. In the rest of this proof, we calculate with the explicit generating sets for submodules given in Theorem 4.4.

Suppose first that $H^2(C,B_1) \cong \mathbb{Z}_2$; equivalently, $(B_1 \cap X)^{1+a} = B_1^{\hat{c}}$. For all such cases except $B_1 = C(i,j,k,0,1,-1)$ we have $B_1^{1-ac} \leq B_1^{1-c}$, implying $E_3^{0,2} \cong H^2(C,B_1)$ by Proposition 3.2.3. Also, in the notation of the proposition, we may choose $m'' \in B_1 \cap XY$ here, so that $d_3^{0,2} = 0$. If $B_1 = C(i,j,k,0,1,-1)$ then $E_3^{0,2} = 0$. Suppose now that $H^2(C,B_1) \cong \mathbb{Z}_4$. Then $(B_1 \cap X)^{1+a} \not\leq B_1^{\hat{c}}$ but $(B_1 \cap X)^{2+2a} = B_1^{\hat{c}} = (B_1 \cap X)^{\hat{c}}$, so that $E_3^{0,2} = (B_1 \cap X)^2/B_1^{\hat{c}} = H^2(C,B_1)^2$ and $d_3^{0,2}$ is trivial again. □

It is now a simple matter to calculate $|H^2(D,B_1)|$ by (3.13), Propositions 5.3–5.5 and Table 5.1. We state the results of this calculation formally in the next theorem.

Theorem 5.6 *Let B_1 be a faithful finite D-submodule of B. Then*

$$|H^2(D,B_1)| = \begin{cases} 16 & B_1 = C(i,j,k,0,\delta_2), \\ 4 & B_1 = C(i,j,k,0,1,-1) \text{ or } B_1 = C(i,j,k,1,0,1), \\ 2 & B_1 = C(i,j,k,1,1,2,\alpha), \\ 8 & \text{otherwise.} \end{cases}$$

By Theorem 2.5 and since $\mathsf{N}_{S_4}(D) = D$, the number of BS_4-conjugacy classes of finite groups G such that $BG = BD$ and $B \cap G = B_1$ is $|H^2(D,B_1)|$. Then by Proposition 5.2 and Theorem 5.6, and employing techniques similar to those in the proof of Theorem 3.2.9, we obtain the following result.

Theorem 5.7 *Let G be a finite irreducible subgroup of BD such that $BG = BD$. Either G is $GL(4)$-conjugate to a group in the list of Theorem 3.3.14, or G is BS_4-conjugate to a unique group in the following list, where $\varepsilon, \eta, \mu, \nu$ range over $\{0,1\}$ and α ranges over $\{-1,1\}$:*

$$\begin{aligned}
&\langle a x_{i+1}^{\varepsilon} u_{j+1}^{\eta} (y_{k+1} v_{k+1})^{\mu}, c x_{i+1}^{\nu}, C(i,j,k,0,0) \rangle, \\
&\langle a x_{i+1}^{\varepsilon} u_{j+1}^{\eta}, c (x_{i+2} v_{k+1})^{\mu}, C(i,j,k,0,0,1) \rangle, \\
&\langle a x_{i+1}^{\varepsilon} y_{k+1}^{\eta} v_{k+1}^{\eta-\mu}, c (x_{i+2} u_{j+2})^{\mu}, C(i,j,k,0,0,-1) \rangle, \\
&\langle a x_{i+1}^{\varepsilon} (u_{j+2} y_{k+1})^{\eta}, c x_{i+1}^{\mu}, C(i,j,k,0,0,0) \rangle,
\end{aligned}$$

$$\begin{aligned}
&\langle a x_{i+1}^{\varepsilon} u_{j+1}^{\eta} y_{k+1}^{\mu}, c x_{i+1}^{\nu}, C(i,j,k,0,1) \rangle, \\
&\langle a x_{i+2}^{\mu-2\varepsilon} u_{j+1}^{\eta} (y_{k+2} v_{k+2})^{\mu}, c x_{i+2}^{\mu}, C(i,j,k,0,1,1) \rangle, \\
&\langle a x_{i+1}^{\varepsilon} u_{j+1}^{\eta}, c, C(i,j,k,0,1,-1) \rangle, \\
&\langle a x_{i+1}^{\varepsilon} u_{j+1}^{\eta}, c x_{i+1}^{\mu}, C(i,j,k,0,1,0) \rangle,
\end{aligned}$$

$$\begin{aligned}
&\langle a x_{i+1}^{\varepsilon} (y_{k+1} v_{k+1})^{\eta}, c (x_{i+2} u_{j+2})^{\mu}, C(i,j,k,1,0) \rangle, \\
&\langle a u_{j+2}^{\varepsilon} y_{k+1}^{\eta-\varepsilon}, c x_{i+2}^{\eta}, C(i,j,k,1,0,1) \rangle,
\end{aligned}$$

$$\begin{aligned}
&\langle a x_{i+1}^{\varepsilon} y_{k+1}^{\eta}, c (x_{i+2} u_{j+2})^{\mu}, C(i,j,k,1,1) \rangle, \\
&\langle a (x_{i+2} y_{k+2} v_{k+2})^{\varepsilon} u_{j+1}^{\eta}, c x_{i+2}^{\mu} u_{j+2}^{\varepsilon-\mu}, C(i,j,k,1,1,1) \rangle, \\
&\langle a x_{i+1}^{\varepsilon}, c, C(i,j,k,1,1,2,\alpha) \rangle.
\end{aligned}$$

If $B \cap G = C(i,j,k,\delta_1,\delta_2)$ then $k \geq 1$, and $i,j \geq 1$ when $\delta_1 = 0$ but $i,j \gtrsim 0$ when $\delta_1 = 1$. If $B \cap G = C(i,j,k,\delta_1,\delta_2,\xi,\alpha)$ is non-Cartesian then i,j,k range over the non-negative integers according to the conditions stated in Theorem 4.4.

Remark 5.8 The diagonal subgroup of each group in the list of Theorem 5.7 is a faithful D-module.

Next, we consider $GL(4)$-conjugacy between groups in the lists of Theorems 3.3.14, 4.8, and 5.7.

Lemma 5.9 *Let G be a finite irreducible subgroup of BD such that $BG = BD$. Then G has class greater than 2.*

Proof. If G had class 2 then $G/B \cap G$ would be abelian. □

Lemma 5.10 *Let G be a group in the list of Theorem 5.7. If G has a self-centralising normal subgroup S of index 4, then $Q = B \cap S$ is a D-submodule of $B \cap G$ of index 2 such that $\mathsf{C}_D(Q) = \langle ac, b \rangle$. In this case, $B \cap G$ is one of the $C(i, j, 0, 0, 1, \xi)$, $C(i, j, 1, \delta_1, 0)$ or $C(i, j, 0, 1, 1, 1)$.*

Proof. The first part is standard (cf. Lemma 3.3.4), after appealing to Remark 5.8 and Lemma 5.9. The second part follows from the first and the comments made after Lemma 4.5. □

Proposition 5.11 *Modify the list of Theorem 5.7 by omitting each group with diagonal subgroup one of the $C(i, j, 0, 0, 1, \xi)$, $C(i, j, 1, \delta_1, 0)$ or $C(i, j, 0, 1, 1, 1)$. Then a finite irreducible subgroup G of BD such that $BG = BD$ is $GL(4)$-conjugate to a group in this modified list or one listed in Theorem 3.3.14 or Theorem 4.8. However, no group in this modified list is isomorphic to a group listed in Theorem 3.3.14 or Theorem 4.8.*

Proof. The second claim is a consequence of Lemma 5.10. So suppose that G is in the list of Theorem 5.7 and proposed for omission. By Proposition 1.3.6, the theorem is proved once we have shown that in all cases G has a normal abelian subgroup S of index 4.

From the list in Theorem 5.7 we can see that $c \in GXU$ and $a \in GXU(y_2v_2)^m$ for some $m \in \{0, 1\}$. By Lemma 5.10, the faithful D-module $B \cap G$ has a (noncyclic) submodule Q of index 2 such that $\mathsf{C}_D(Q) = \langle ac, b \rangle$. Choose $w \in (B \cap G) \backslash Q$. Then we have in G a subgroup of the form $S = \langle acz_{ac}(y_2v_2)^m, bx_bw^m, Q \rangle$, where $z_{ac} \in XU$ and $x_b \in X$. It is easy to see that S is abelian and has index 4 in G. Certainly S is normal in $(B \cap G)S$, since $(B \cap G)/Q$ is a normal subgroup of order 2 in G/Q and hence $(B \cap G)/Q \leq Z(G/Q)$. Therefore, to prove normality of S in G it is enough to prove that S is closed under conjugation by a single element g of $G \backslash (B \cap G)S$. For instance, if $g \in cB \cap G$ then $S^g = S$ provided that $x_b(wy_1)^m z_{ac}^{1-c} \in Q$: a condition readily verified by inspection of the list in Theorem 5.7. □

It remains to determine $GL(4)$-conjugacy between groups in the modified list of Proposition 5.11. We will outline below a strategy for doing this (a variant of the one used in Section 3.3), but first we determine the candidates for diagonal subgroups of conjugate groups.

Suppose that G and H are groups in the list such that $G^m = H$ for some $m \in GL(4)$. Then necessarily $Q = B \cap G \cap (B \cap H)^{m^{-1}}$ is a D-submodule of $B \cap G$ of index 4, and is maximal with respect to the condition $\mathsf{C}_D(Q) = \langle ac, b \rangle$. Also, $Q^m = Q$. If

$B\cap G$ is Cartesian (with respect to (4.1)) then $(B\cap G)/F_b$ splits over $Q/F_b \leq XU/F_b$, by maximality. For fixed Q there is only one choice for the complement of order 4, since YV/F_b is uniserial. The corresponding choices for $B\cap G$ are $C(i,j,1,0,1)$ or $C(i,j,1,1,1)$ for some $i,j \gtrsim 0$. Suppose now that $B\cap G$ is non-Cartesian. Then Q is contained in the unique maximal Cartesian submodule M of $B\cap G$, and hence $M=Q$ or $|M:Q|=2$. In the first case clearly $B\cap G$ is one of the $C(i,j,0,1,1,2,\alpha)$, for some $i,j\gtrsim 0$. In the second case, M is a faithful Cartesian submodule in a page of the form $\mathcal{L}(C(i+1,j+1,1,0,0)/C(i,j,0,0,1))$. Thus M is $C(i,j,1,1,0)$ or $C(i,j,1,0,0)$ for some i,j (see Figure 4.1). Fixing M fixes the first five parameters in the label for $B\cap G$, and consequently $B\cap G$ is $C(i,j,1,1,0,1)$ or one of the $C(i,j,1,0,0,\xi)$.

The condition $Q^m = Q$ implies that, up to postmultiplication by an element of BS_4, m acts as conjugation by $a^\sigma wf$, where $\sigma\in\{0,1\}$, w is a suitably chosen element of YV, and f is the $(2,3)$-conjugate of a matrix of the form (3.21). So the first step in our strategy is to calculate $G^{a^\sigma wf}$ for $\sigma=0,1$ and each G in the list of Proposition 5.11 such that $B\cap G$ is one of the submodules determined in the previous paragraph. If neither G^{wf} nor G^{awf} lie in BD then G cannot be conjugate to another group in the list of Proposition 5.11. If $G^{a^\sigma wf}\leq BD$, then the second step in our strategy is to select the unique BD-conjugate H of $G^{a^\sigma wf}$ in the list of Proposition 5.11. In this way, all possible conjugacies $G\sim_{GL(4)} H$ are determined. These will not be explicitly stated, but are used in formulation and verification of the following theorem. This completes our solution of the listing problem for $T=D$ and thence overall.

Theorem 5.12 *Modify the list of Proposition* 5.11 *by omitting each group with diagonal subgroup one of the* $C(i,j,0,1,1,2,\alpha)$, $i,j\gtrsim 0$, *and also omitting*:

$\langle ax_{i+1}^{\varepsilon}u_{j+1}^{\eta}, cx_{i+1}^{\nu}, C(i,j,1,0,1)\rangle$, $i\geq 2$ *or* $j\geq 2$;
$\langle ax_2u_2^{\eta}, cx_2^{\nu}, C(1,1,1,0,1)\rangle$;
$\langle au_2^{\eta}, c, C(1,1,1,0,1)\rangle$;
$\langle a, c, C(i,j,1,1,0,1)\rangle$, $i,j\geq 1$;
$\langle au_2, cx_2, C(0,0,1,1,0,1)\rangle$;
$\langle ax_{i+1}^{\varepsilon}u_{j+1}, c, C(i,j,1,0,0,1)\rangle$, $i\geq 1$ *and* $j\geq 2$;
$\langle ax_{i+1}^{\varepsilon}, c, C(i,1,1,0,0,1)\rangle$, $i\geq 1$;
$\langle ax_{i+1}^{\varepsilon}, cx_{i+1}^{1-\varepsilon}, C(i,j,1,0,0,0)\rangle$, $i\geq 2$ *and* $j\geq 1$;
$\langle ax_2^{\varepsilon}, cx_2^{\varepsilon}, C(1,j,1,0,0,0)\rangle$, $j\geq 1$.

Then this modified list is a complete and irredundant list of $GL(4)$*-conjugacy class representatives of the finite irreducible subgroups* G *of* BD *such that* $BG=BD$. *No group in this list is isomorphic to a group in the list of Theorem* 3.3.14 *or Theorem* 4.8.

Chapter 6

Full solutions

In this chapter we carry out two tasks. First, by way of summarising and consolidating results obtained to this point, we present in its entirety our solution of the main listing problem. Secondly, we answer the restricted isomorphism question for finite irreducible 2-subgroups of $GL(4)$, as posed in Chapter 2.

6.1 The final list

Theorem 6.1.1 *The distinct groups defined by the generating sets below, where the parameters range independently as indicated, constitute a complete and irredundant list of conjugacy class representatives of the finite irreducible 2-subgroups of $GL(4)$.*

(i) *For $\varepsilon, \eta, \gamma, \mu, \nu \in \{0,1\}$, $\alpha \in \{-1,1\}$ and $i,j,k,l \geq 1$:*

$$\begin{array}{ll}
\langle ax_{i+1}^{\varepsilon} y_{j+1}^{\eta}, bx_{i+1}^{\gamma} u_{k+1}^{\mu} v_{l+1}^{\nu}, F(i,j,k,l,0,0)\rangle, & j < k < l; \\
\langle ay_{j+1}^{\eta}, bx_{i+1}^{\gamma} u_{k+1}^{\mu} v_{k+1}^{\nu\mu}, F(i,j,k,k,0,0)\rangle, & j \neq k; \\
\langle ax_{i+1} y_{j+1}^{\eta}, bu_{k+1}^{\mu} v_{k+1}^{\nu}, F(i,j,k,k,0,0)\rangle, & j \neq k; \\
\langle ay_{j+1}^{\mu}, bu_{j+1}^{\mu} v_{j+1}^{\nu(1-\mu)+1}, F(i,j,j,j,0,0)\rangle; & \\
\langle ax_{i+1} y_{j+1}^{\eta}, bu_{j+1}^{\mu} v_{j+1}^{\nu\eta}, F(i,j,j,j,0,0)\rangle; & \\
\langle ay_{j+1}^{\eta}, bx_{i+1}^{\gamma} u_{k+1}^{\mu}, F(i,j,k,l,0,0,1,-1)\rangle, & k < l; \\
\langle ay_{j+1}^{\eta}, bx_{i+1}^{\gamma(1-\mu)} u_{k+1}^{\mu}, F(i,j,k,k,0,0,1,-1)\rangle; & \\
\langle ax_{i+1}^{\varepsilon}, bx_{i+1}^{\gamma} u_{k+1}^{\mu}, F(i,j,k,l,0,0,0,-1)\rangle, & j < k < l; \\
\langle ax_{i+1}^{\varepsilon}, bx_{i+1}^{\gamma(1-\varepsilon)} u_{k+1}^{\mu}, F(i,j,k,k,0,0,0,-1)\rangle, & j \neq k; \\
\langle ax_{i+1}^{\varepsilon}, bu_{j+1}^{\mu}, F(i,j,j,j,0,0,0,-1)\rangle; & \\
\langle ax_{i+1}^{\varepsilon}, bx_{i+1}^{\gamma} u_{k+1}^{\mu}, F(i,j,k,l,0,0,-1,-1)\rangle, & j < k < l; \\
\langle ax_{i+1}^{\varepsilon}, bx_{i+1}^{\gamma(1-\varepsilon-\mu)} u_{k+1}^{\mu}, F(i,j,k,k,0,0,-1,-1)\rangle, & j \neq k; \\
\langle ax_{i+1}^{\varepsilon}, bu_{j+1}^{\mu}, F(i,j,j,j,0,0,-1,-1)\rangle; &
\end{array}$$

$i,j \geq 1,\ k,l \gtrsim 0:$

$$\begin{aligned}
&\langle ax_{i+1}^{\varepsilon}y_{j+1}^{\eta}, bx_{i+1}^{\gamma}, F(i,j,k,l,0,1)\rangle, \quad k<l;\\
&\langle ax_{i+1}^{\varepsilon}y_{j+1}^{\eta}, bx_{i+1}^{\gamma(1-\varepsilon)}, F(i,j,k,k,0,1)\rangle;\\
&\langle ay_{j+1}^{\eta}, b(x_{i+2}u_2v_2)^{1-\gamma\eta}, F(i,j,0,0,0,1,1)\rangle;
\end{aligned}$$

$i,j,k,l \geq 1:$

$$\begin{aligned}
&\langle ax_{i+1}^{\varepsilon}, b, F(i,j,k,l,0,1,-1)\rangle, \quad k \leq l;\\
&\langle ax_{i+1}^{\varepsilon}, bx_{i+1}^{\gamma}, F(i,j,k,l,0,1,0)\rangle, \quad j<k<l;\\
&\langle ax_{i+1}^{\varepsilon}, bx_{i+1}^{\gamma(1-\varepsilon)}, F(i,j,k,k,0,1,0)\rangle, \quad j \neq k;\\
&\langle ax_{i+1}^{\varepsilon}, b, F(i,j,j,j,0,1,0)\rangle;
\end{aligned}$$

$i,j \gtrsim 0,\ k,l \geq 1:$

$$\begin{aligned}
&\langle ax_{i+1}^{\varepsilon}, bu_{k+1}^{\mu}v_{l+1}^{\nu}, F(i,j,k,l,1,0)\rangle, \quad k<l;\\
&\langle ax_{i+1}^{\varepsilon}, bu_{k+1}^{\mu}v_{k+1}^{\nu\mu}, F(i,j,k,k,1,0)\rangle;\\
&\langle a(x_{i+2}y_{j+2})^{\varepsilon}, b(x_{i+2}u_{k+2})^{\varepsilon}v_{l+1}^{\nu}, F(i,j,k,l,1,0,1)\rangle, \quad j \leq k;\\
&\langle a, bv_{l+1}^{\nu}, F(i,j,k,l,1,0,-1)\rangle, \quad k \leq l;
\end{aligned}$$

$i,j \gtrsim 0,\ k,l \gtrsim 0:$

$$\begin{aligned}
&\langle ax_{i+1}^{\varepsilon}, b, F(i,j,k,l,1,1)\rangle, \quad k \leq l;\\
&\langle a(x_{i+2}y_{j+2})^{\varepsilon}, b(x_{i+2}u_{k+2})^{\gamma}v_{l+2}^{\gamma-\varepsilon}, F(i,j,k,l,1,1,1)\rangle, \quad j<k<l;\\
&\langle a(x_{i+2}y_{j+2})^{\varepsilon}, b(x_{i+2}u_2)^{1-\varepsilon}v_2, F(i,j,0,0,1,1,1)\rangle, \quad j \geq 1;\\
&\langle a(x_{i+2}y_{j+2})^{\varepsilon\gamma}, b(x_{i+2}u_{k+2})^{\gamma(1-\varepsilon)}v_{k+2}^{\gamma}, F(i,j,k,k,1,1,1)\rangle, \quad k \geq 1,\ j \neq k;\\
&\langle a, bx_2u_2v_2, F(0,0,0,0,1,1,1)\rangle;\\
&\langle a, b(x_{i+2}u_{j+2}v_{j+2})^{\gamma}, F(i,j,j,j,1,1,1)\rangle, \quad j \geq 1;\\
&\langle a, b, F(0,0,k,l,1,1,2,\alpha)\rangle, \quad k \leq l;\\
&\langle a, b, F(i,j,k,l,1,1,2,\alpha)\rangle, \quad i \geq 1,\ j \leq k \leq l.
\end{aligned}$$

(ii) *For* $\varepsilon \in \{0,1,2\}$, $\eta,\delta \in \{0,1\}$ *and* $\xi,\alpha \in \{-1,1\}$:

$$\left.\begin{array}{ll}
\langle cx_{i+2}^{\varepsilon}, C(i,j,k,0,0)\rangle, & k \geq 2\\
\left.\begin{array}{l}
\langle cx_{i+2}^{\varepsilon}, C(i,j,k,0,1)\rangle\\
\langle cx_{i+2}^{\varepsilon}, C(i,j,k,0,\delta,0)\rangle\\
\langle cx_{i+2}^{\eta}, C(i,j,k,0,\delta,\xi)\rangle
\end{array}\right\} & k \geq 1
\end{array}\right\} i,j \geq 1;$$

$$\left.\begin{array}{ll}
\langle cx_{i+2}^{\eta}, C(i,j,k,1,0)\rangle, & k \geq 2\\
\left.\begin{array}{l}
\langle cx_{i+2}^{\eta}, C(i,j,k,1,1)\rangle\\
\langle cx_{i+2}^{\eta}, C(i,j,k,1,\delta,1)\rangle\\
\langle c, C(i,j,k,1,0,2,1)\rangle
\end{array}\right\} & k \geq 1
\end{array}\right\} i,j \gtrsim 0;$$

$$\langle c, C(i,j,k,1,1,2,\alpha)\rangle, \quad i,j \gtrsim 0,\ k \ge 0.$$

(iii) *For* $\varepsilon, \eta, \mu, \nu \in \{0,1\}$ *and* $\alpha \in \{-1,1\}$:

$$\begin{array}{l}
\left.\begin{array}{l}
\langle a x_{i+1}^{\varepsilon} u_{j+1}^{\eta} (y_{k+1}v_{k+1})^{\mu}, c x_{i+1}^{\nu}, C(i,j,k,0,0)\rangle \\
\langle a x_{i+1}^{\varepsilon} u_{j+1}^{\eta}, c(x_{i+2}v_{k+1})^{\mu}, C(i,j,k,0,0,1)\rangle \\
\langle a x_{i+1}^{\varepsilon} (u_{j+2}y_{k+1})^{\eta}, c x_{i+1}^{\mu}, C(i,j,k,0,0,0)\rangle,
\end{array}\right\} \ i,j \ge 1, k \ge 2; \\
\langle a x_{i+1}^{\varepsilon} y_{k+1}^{\eta} v_{k+1}^{\eta-\mu}, c(x_{i+2}u_{j+2})^{\mu}, C(i,j,k,0,0,-1)\rangle, \quad i,j,k \ge 1; \\
\left.\begin{array}{l}
\langle a x_{i+1}^{\varepsilon} u_{2}^{\eta}, c(x_{i+2}v_{2})^{1-\eta\mu}, C(i,1,1,0,0,1)\rangle \\
\langle a x_{i+1}^{\varepsilon} u_{j+1}^{\eta\mu}, c(x_{i+2}v_{2})^{\mu}, C(i,j,1,0,0,1)\rangle, \quad j \ge 2
\end{array}\right\} \ i \ge 1; \\
\left.\begin{array}{l}
\langle a x_{2}^{\varepsilon}, c x_{2}^{1-\varepsilon}, C(1,j,1,0,0,0)\rangle \\
\langle a x_{i+1}^{\varepsilon}, c x_{i+1}^{\varepsilon}, C(i,j,1,0,0,0)\rangle, \quad i \ge 2 \\
\langle a x_{i+1}^{\varepsilon} u_{j+2} y_{2}, c x_{i+1}^{\eta}, C(i,j,1,0,0,0)\rangle, \quad i \ge 1
\end{array}\right\} \ j \ge 1;
\end{array}$$

$$\begin{array}{l}
\langle a u_{2}^{\eta}, c x_{2}, C(1,1,1,0,1)\rangle; \\
\left.\begin{array}{l}
\langle a x_{i+1}^{\varepsilon} u_{j+1}^{\eta} y_{2}, c x_{i+1}^{\nu}, C(i,j,1,0,1)\rangle \\
\langle a x_{i+1}^{\varepsilon} u_{j+1}^{\eta} y_{k+1}^{\mu}, c x_{i+1}^{\nu}, C(i,j,k,0,1)\rangle, \quad k \ge 2 \\
\left.\begin{array}{l}
\langle a x_{i+2}^{\mu-2\varepsilon} u_{j+1}^{\eta} (y_{k+2}v_{k+2})^{\mu}, c x_{i+2}^{\mu}, C(i,j,k,0,1,1)\rangle \\
\langle a x_{i+1}^{\varepsilon} u_{j+1}^{\eta}, c, C(i,j,k,0,1,-1)\rangle \\
\langle a x_{i+1}^{\varepsilon} u_{j+1}^{\eta}, c x_{i+1}^{\mu}, C(i,j,k,0,1,0)\rangle
\end{array}\right\} \ k \ge 1
\end{array}\right\} \ i,j \ge 1;
\end{array}$$

$$\begin{array}{l}
\langle a, c, C(0,0,1,1,0,1)\rangle; \\
\langle a u_{j+2}, c x_{i+2}, C(i,j,1,1,0,1)\rangle, \quad i,j \ge 1; \\
\left.\begin{array}{l}
\langle a u_{j+2}^{\varepsilon} y_{2}, c x_{i+2}^{1-\varepsilon}, C(i,j,1,1,0,1)\rangle \\
\left.\begin{array}{l}
\langle a x_{i+1}^{\varepsilon} (y_{k+1}v_{k+1})^{\eta}, c(x_{i+2}u_{j+2})^{\mu}, C(i,j,k,1,0)\rangle \\
\langle a u_{j+2}^{\varepsilon} y_{k+1}^{\eta-\varepsilon}, c x_{i+2}^{\eta}, C(i,j,k,1,0,1)\rangle
\end{array}\right\} \ k \ge 2
\end{array}\right\} \ i,j \gtrsim 0;
\end{array}$$

$$\left.\begin{array}{l}
\langle a x_{i+1}^{\varepsilon} y_{k+1}^{\eta}, c(x_{i+2}u_{j+2})^{\mu}, C(i,j,k,1,1)\rangle \\
\langle a(x_{i+2}y_{k+2}v_{k+2})^{\varepsilon} u_{j+1}^{\eta}, c x_{i+2}^{\mu} u_{j+2}^{\varepsilon-\mu}, C(i,j,k,1,1,1)\rangle \\
\langle a x_{i+1}^{\varepsilon}, c, C(i,j,k,1,1,2,\alpha)\rangle
\end{array}\right\} \ i,j \gtrsim 0,\ k \ge 1.$$

The list in Theorem 6.1.1 is the union of three sublists: (i) the groups in BV_4; (ii) the groups in BC; (iii) the supplements of B in BD. Theorems 3.3.1 and 3.3.14 furnish sublist (i). Sublist (ii) is just a restatement of Theorem 4.8. Sublist (iii) is obtained from Theorem 5.7, Proposition 5.11 and Theorem 5.12. The completeness and irredundancy claim in Theorem 6.1.1 is also evident from these stated results. Note that groups in different sublists are not isomorphic.

6.2 The restricted isomorphism question

Throughout, T is a transitive 2-subgroup of S_4.

Lemma 6.2.1 *Let $B_1 \neq 1$ be a faithful finite T-submodule of B and let λ be a T-automorphism of B_1.*

(i) *If $|T| = 4$ then λ is induced by the action of some $\sum_{t\in T} \lambda_t t \in \mathbb{Z}T$, where the number of coefficients λ_t that are odd is one or three.*

(ii) *If $T = D$ then λ is induced by the action of some $\sum_{t\in V_4} \lambda_t t \in \mathbb{Z}V_4$, where $\lambda_a - \lambda_{ab}$ is even and the number of coefficients λ_t that are odd is one or three.*

Proof. (i) Recall that we may identify B_1 with a submodule of the regular module of a group algebra over T. The first part of the assertion is then implicit in the proof of Proposition 1.1.2. If the second part were not true then $x_0 \in B_1$ would be mapped to 1 under λ.

(ii) Two of the claims are immediate from (i) and the observation that λ is a V_4-automorphism of B_1. Also, λ commutes with (conjugation by) c. This implies that $\lambda_a - \lambda_{ab}$ is even: otherwise, B_1 would be $\langle b\rangle$-trivial, since odd-powering is an automorphism of B_1. □

Definition 6.2.2 Let G be a finite subgroup of BT such that $B \cap G$ is a faithful T-module, and let π be projection of G onto T. Choose an arbitrary T-automorphism of $B \cap G$, acting as $\lambda \in \mathbb{Z}T$, by Lemma 6.2.1. Then we denote the finite subgroup

$$\{g\pi((g\pi)^{-1}g)^{\lambda} \mid g \in G\}$$

of BT by G_λ. Of course, $B \cap G_\lambda = B \cap G_{\lambda'}$ for any two T-automorphisms λ, λ' of $B \cap G$. Note also that $g \mapsto g\pi((g\pi)^{-1}g)^{\lambda}$, $g \in G$, defines an isomorphism of G onto G_λ.

Proposition 6.2.3 *Let $T = V_4$ or D and suppose that G is one of the groups listed in Theorem 3.3.1 or Theorem 5.7. If λ is a T-automorphism of $B\cap G$ then $G_\lambda \sim_{BT} G$.*

Proof. Let $\equiv$ denote equality modulo $B \cap G$, unless stated otherwise.

For each $g \in G$, $((g\pi)^{-1}g)^{1+a+b+ab} \in G \cap X$. Write λ as $\sum_{t\in T} \lambda_t t \in \mathbb{Z}T$ and set $\lambda' = \lambda + 1 + a + b + ab$. If precisely three of the λ_t are odd, then

$$g\pi((g\pi)^{-1}g)^{\lambda} \equiv g\pi((g\pi)^{-1}g)^{\lambda}((g\pi)^{-1}g)^{1+a+b+ab} \equiv g\pi((g\pi)^{-1}g)^{\lambda'},$$

showing that $G_\lambda = G_{\lambda'}$. By Lemma 6.2.1 (and Remark 5.8), we may therefore assume that precisely one of the λ_t is odd; call it λ_r.

Suppose that $T = V_4$. If $g\pi = a$ then $(ag)^2 \in B \cap G$, so that $(ag)^\lambda \equiv (ag)^r$. If $g\pi = b$ then inspection of the list in Theorem 3.3.1 reveals that $(bg)^2 \in B \cap G$ except possibly when $B \cap G = F(i, j, k, l, 0, 1, 1)$ for $k = l = 0$, or when $B \cap G = F(i, j, k, l, 1, 1, 1)$. In those cases, $(b(bg)^\lambda)^{u_{k+2}v_{l+2}} \equiv bx_{i+1}u_{k+1}v_{l+1}(bg)^r v_{l+1} \equiv b(bg)^r$. All of this shows that either $G_\lambda = G^r$ or $(G_\lambda)^{u_{k+2}v_{l+2}} = G^r$ for some k and l, proving the proposition for $T = V_4$.

Suppose now that $T = D$. In this case, $r = 1$ or $r = b$, since $\lambda_a - \lambda_{ab}$ is even by Lemma 6.2.1. Choose $g \in G$ and write $(g\pi)^{-1}g$ as $xyuv$ for $x \in X$, $y \in Y$, $u \in U$ and $v \in V$. Since $y^4 \equiv 1$ always, we have $y^{\lambda - r} = y^{\lambda_1' + \lambda_a - \lambda_b' - \lambda_{ab}} \equiv y^{\lambda_1' + \lambda_a + \lambda_b' + \lambda_{ab}}$, where λ_1' and λ_b' are even. Similar statements hold when y is replaced by u and then v, so that

$$g\pi((g\pi)^{-1}g)^\lambda \equiv g\pi((g\pi)^{-1}g)^{r+n} = g^r((g\pi)^{-1}g)^n,$$

where $n = 0$ or 2. If $n = 0$ for all $g \in G$ then $G_\lambda = G^r$. The rest of the proof is a case-by-case verification, with the aid of the list in Theorem 5.7, that $G_\lambda \sim_{BT} G^r$ even when $n = 2$ for some $g \in G$. The manipulations are straightforward and similar to those performed above for $T = V_4$; details are omitted. □

We are now able to prove that the restricted isomorphism question is answered affirmatively.

Theorem 6.2.4 *Let G and H be finite irreducible 2-subgroups of BS_4. If there is an isomorphism θ of G onto H such that $(B \cap G)\theta = B \cap H$, then G and H are conjugate in $GL(4)$.*

Proof. Since $G/B \cap G \cong H/B \cap H$, after any necessary S_4-conjugation we have $BG = BH = BT$, for T one of V_4, C or D. Recall the conventions of Chapter 2.

Initially suppose that G and H are both listed in Theorem 3.3.1, Theorem 4.8, or Theorem 5.7. In the second of these cases $G = H$, so we consider only the first and third. By Theorem 2.7 and Remark 3.3.2, $B \cap G = B \cap H = B_1$, say. Denote by $[\psi_1]$ and $[\psi_2]$ the elements of $H^2(T, B_1)$ corresponding to the equivalence classes of the extensions $0 \to B_1 \stackrel{\text{inc.}}{\to} G \stackrel{\text{proj.}}{\to} T \to 1$ and $0 \to B_1 \stackrel{\text{inc.}}{\to} H \stackrel{\text{proj.}}{\to} T \to 1$, respectively. By the discussion framing (2.3), ψ_2 is cohomologous to $\psi_1^{(\theta_{B_1}, s)}$ for some $s \in \mathsf{N}_{S_4}(T)$; in fact, compatibility and Proposition 1.1.2 force $B_1^s = B_1$ (this is the argument in the first few lines in the proof of Theorem 2.7). Choose a transversal function $\sigma: T \to G$ for $0 \to B_1 \stackrel{\text{inc.}}{\to} G \stackrel{\text{proj.}}{\to} T \to 1$. Then $\tilde{\sigma}: T \to G^s$, defined by $t\tilde{\sigma} = (t^{s^{-1}}\sigma)^s$, is a transversal function for

$$0 \to B_1 \stackrel{\lambda}{\to} G^s \stackrel{\text{proj.}}{\to} T \to 1, \tag{6.1}$$

where $\lambda = \theta_{B_1}^{-1}s$ (here we are using implicitly the fact that λ is a T-automorphism of B_1). The 2-cocycle arising from (6.1) for this choice of transversal function is defined by $(t_1, t_2) \mapsto ((t_1^{s^{-1}}t_2^{s^{-1}}\sigma)^{-1}.t_1^{s^{-1}}\sigma.t_2^{s^{-1}}\sigma)\theta_{B_1}$, and is therefore cohomologous to $\psi_1^{(\theta_{B_1}, s)}$.

Now λ and $s\lambda^{-1}s^{-1}$ are T-automorphisms of B_1. So it follows from Proposition 6.2.3 that $G_{s\lambda^{-1}s^{-1}} \sim_{BT} G$, whence $(G^s)_{\lambda^{-1}} \sim_{BT} G^s$. But

$$0 \to B_1 \overset{\text{inc.}}{\to} (G^s)_{\lambda^{-1}} \overset{\text{proj.}}{\to} T \to 1 \tag{6.2}$$

is equivalent to (6.1); for instance, the isomorphism of G^s onto $(G^s)_{\lambda^{-1}}$ noted in Definition 6.2.2 is an equivalence. Thus (6.2) and $0 \to B_1 \overset{\text{inc.}}{\to} H \overset{\text{proj.}}{\to} T \to 1$ are equivalent, implying that $G^s \sim_{BT} (G^s)_{\lambda^{-1}} \sim_{BT} H$ by Proposition 2.4. Since distinct groups listed in Theorem 3.3.1 and Theorem 5.7 are not BS_4-conjugate, we have proved that $G = H$.

More generally, suppose that G and H are BS_4-conjugate to groups $\overline{G}$ and $\overline{H}$ listed in Theorem 3.3.1, Theorem 4.8, or Theorem 5.7. There is certainly an isomorphism of $\overline{G}$ onto $\overline{H}$ mapping $B \cap \overline{G}$ onto $B \cap \overline{H}$. Hence $\overline{G} = \overline{H}$ by the previous paragraph, and so $G \sim_{BS_4} H$.

The only possibility not covered so far is that G has an SCN4 subgroup A such that G^x is in the list of Theorem 3.3.1 with $B \cap G^x = A^x$, for some $x \in GL(4)$. By Proposition 1.3.6, there is $y \in GL(4)$ such that $H^y \leq BV_4$ and $B \cap H^y = A\theta^y$. Since $A\theta$ is noncyclic, H^y is BS_4-conjugate to a group listed in Theorem 3.3.1. There is an isomorphism between G^x and this group that preserves diagonal subgroups (namely $x^{-1}\theta y$, up to BS_4-conjugation). Appealing to the second paragraph once more, we get $G \sim H$ as required. □

Chapter 7

Schur indices

Let G be a finite irreducible 2-subgroup of $GL(4)$, and denote by χ the trace map on G (the character of the defining representation of G). Fix a subfield $\mathbb{F}$ of $\mathbb{C}$ and denote the Schur index of χ over $\mathbb{F}$ by $m(\chi)$. In this chapter we show how to find $m(\chi)$ for given G. Our approach is to solve the equivalent problem for a maximal subgroup of G and a character of that subgroup with smaller degree than χ.

A reference for the theory of Schur indices is Chapter 10 of [13], the notation and basic results of which we will adopt. I am indebted to Dr L.G. Kovács for discussions concerning some of the proofs below.

From now on, we assume without loss of generality that $G \leq BT$, where $T = V_4$, C or D. Since T has an intransitive subgroup of index 2, G has a reducible subgroup H of index 2 (of course, H is nonabelian). So for some characters ζ and ζ' of H, where ζ is irreducible of degree 2, we have $\chi_H = \zeta + \zeta'$. Frobenius reciprocity and comparison of character degrees show that $\zeta' \neq \zeta$ is also irreducible of degree 2, and $\chi = \zeta^G = (\zeta')^G$. Clearly, then, $\mathbb{F}(\chi) \subseteq \mathbb{F}(\zeta)$. Furthermore, if $h \in H$ and $g \in G \backslash H$ then

$$\zeta(h) + \zeta'(h) = \chi(h) = \zeta^G(h) = \zeta(h) + \zeta(h^g),$$

so that $\mathbb{F}(\zeta) = \mathbb{F}(\zeta')$. The next few results will establish that finding $m(\chi)$ is the same as finding $m(\zeta)$, for an irreducible character ζ of a particular H maximal in G.

Lemma 7.1 *$m(\zeta)$ divides $m(\chi)$.*

Proof. See Problem 10.1 (b), p.171 of [13]. □

Lemma 7.2 *If $\mathbb{F}(\chi) = \mathbb{F}(\zeta)$ then $m(\chi) = m(\zeta)$.*

Proof. This follows from Lemma 10.4, p.162 of [13], and Lemma 7.1. □

Proposition 7.3 *Either $m(\chi) = m(\zeta)$, or $\mathbb{F}(\chi) \neq \mathbb{F}(\zeta)$ and $\{\zeta, \zeta'\}$ is a Galois conjugacy class over $\mathbb{F}(\chi)$.*

Proof. First note that $m(\chi), m(\zeta) \in \{1,2\}$ by Corollary 10.14, p.168 of [13]. By Lemmas 7.1 and 7.2, we may therefore assume that $m(\chi) = 2$ and $m(\zeta) = 1$ in this proof.

By assumption, there is an irreducible $\mathbb{F}(\chi)$-representation $\mathcal{R}$ of G affording 2χ. Suppose that $\mathcal{R}_H$ is irreducible. Since $\mathcal{R}_H^{\mathbb{C}}$ is the direct sum of two copies of an irreducible $\mathbb{C}$-representation with character ζ and two copies of an irreducible $\mathbb{C}$-representation with character ζ', it follows from Theorem 9.21 (c), p.154 of [13] that $\{\zeta, \zeta'\}$ is a Galois conjugacy class over $\mathbb{F}(\chi)$.

Suppose next that $\mathcal{R}_H$ is reducible. This means that $\mathcal{R}_H = \mathcal{S}_1 + \mathcal{S}_2$ for irreducible $\mathbb{F}(\chi)$-representations such that $\mathcal{S}_1^G = \mathcal{S}_2^G = \mathcal{R}$. Denote by σ_i the character afforded by $\mathcal{S}_i$. Since $m(\zeta) = 1$ and $\mathbb{F}(\zeta) = \mathbb{F}(\zeta')$, both ζ and ζ' are irreducible $\mathbb{F}(\zeta)$-characters. Equating the two expressions $\sigma_1 + \sigma_2$ and $2\zeta + 2\zeta'$ of $2\chi_H$ as a sum of $\mathbb{F}(\zeta)$-characters, we see that either $\sigma_i = 2\zeta$ for some i, or $\sigma_1 = \sigma_2 = \zeta + \zeta'$. The first possibility yields $\mathbb{F}(\chi) = \mathbb{F}(\zeta)$ and then $m(\chi) = m(\zeta)$ by Lemma 7.2, contradicting our initial assumption. The second possibility gives the required conclusion, again by Theorem 9.21(c), p.154 of [13]. □

To compute Schur indices we need to make explicit the reducible subgroup H of G. We can take $B \cap G$ noncyclic without loss. Up to S_4-conjugacy, $B \cap H \geq F_a$, and the elements of H have the form

$$h = \begin{pmatrix} h_1 & 0 \\ 0 & h_2 \end{pmatrix} \tag{7.1}$$

where $h_1, h_2 \in GL(2)$. When $T = V_4$, H is the maximal subgroup of G such that $B \cap H = B \cap G$ and $BH = B\langle a \rangle$. When $T = C$ or $T = D$, H is the $(2,3)$-conjugate of the maximal subgroup M of G such that $B \cap M = B \cap G$ and $BM \leq B\langle ac, b \rangle$.

Corollary 7.4 *Let G be a group in the list of Theorem 6.1.1 and let H be the maximal subgroup of G in the form described above (up to S_4-conjugacy). Then there is an irreducible character ζ of H of degree 2 such that $m(\chi) = m(\zeta)$.*

Proof. Let ζ be the character afforded by the representation mapping $h \in H$ in the form (7.1) to h_1, and let ζ' be the character afforded by the representation mapping h to h_2. Both ζ and ζ' are irreducible. We have $\zeta(x_0) = -2 = \zeta'(x_0)$, yet $\zeta(x_1 y_1) = -2 = -\zeta'(x_1 y_1)$. Thus ζ and ζ' are not Galois conjugate, and the result is a consequence of Proposition 7.3. □

Henceforth, we take $\mathbb{F} = \mathbb{Q}$.

The relevant special case of (14.3), p.73 of [8] is an implicit version of Corollary 7.4, but with no instruction on how to recognise H in G, at least for the purpose of computing $m(\chi)$. However, we have here an explicit description of H, and so can apply to it the following result.

Lemma 7.5 *Let G be a finite irreducible 2-subgroup of $GL(2)$ with $|Z(G)| = 2$ and let ϱ be the trace map on G. Then $m(\varrho) = 2$ if and only if G is generalised quaternion; otherwise, $m(\varrho) = 1$.*

Proof. As usual, $m(\varrho) \in \{1, 2\}$. The hypotheses imply that G is one of the types dihedral, generalised quaternion or semidihedral (see, for example, 5.1 of [5], where a generating set for a $GL(2)$-conjugate of G is given in each of these cases).

If G is dihedral then $m(\varrho) = 1$, and if G is generalised quaternion then $m(\varrho) = 2$, by (11.7) and (11.8), p.64 of [8]. Assume now that $G \cong SD_{2^{k+2}}$, the semidihedral group of order 2^{k+2}, $k \geq 1$. Up to conjugacy, $G = \langle t, d \rangle$, where

$$t = \begin{pmatrix} 0 & 1 \\ 1 & 0 \end{pmatrix}, \qquad d = \sqrt{-1}(\omega_k, \omega_k^{-1}).$$

Now d is a cyclic linear transformation whose minimal polynomial is $f = x^2 - \nu_k x - 1$, where $\nu_k = \varrho(d)$. Thus d is conjugate to the companion matrix m_k of f: $d^e = m_k$, where

$$e = \begin{pmatrix} 1 & \sqrt{-1}\omega_k \\ 1 & \sqrt{-1}\omega_k^{-1} \end{pmatrix}, \qquad m_k = \begin{pmatrix} 0 & 1 \\ 1 & \nu_k \end{pmatrix}.$$

It may be checked that t^e has entries in $\mathbb{Q}(\varrho)$. So the $\mathbb{Q}(\varrho)$-representation of G induced by e affords ϱ, and the proof of the lemma is complete (an argument similar to that given above can also be used to prove directly that $m(\varrho) = 1$ when G is dihedral). □

We now outline a procedure for determining $m(\chi)$ when G is in the list of Theorem 6.1.1. If $|Z(G)| \geq 4$ then G contains the scalar matrix $\sqrt{-1}$; thus $\sqrt{-1} \in \mathbb{Q}(\chi)$ and $m(\chi) = 1$ (see, for example, Corollary 10.14, p.168 of [13]). Those groups for which $|Z(G)| = 2$ are readily determined from the list in Theorem 6.1.1: in the label of $B \cap G$, each of the first two parameters is 0 (if this is admissible). For any such group we apply Corollary 7.4, in conjunction with Lemma 7.5, after any necessary S_4-conjugation of G to obtain H in the required form (7.1). Let $\tilde{H}$ be the image of H under the representation affording ζ (where this character is chosen as in the proof of Corollary 7.4). If $|Z(\tilde{H})| > 2$ then $m(\chi) = 1$; otherwise, $m(\chi) = 2$ or 1 according to whether $\tilde{H}$ is generalised quaternion or not, by Lemma 7.5. For example, after following this procedure for the groups in sublist (ii) of Theorem 6.1.1, we see that only the non-split extensions $\langle cx_2, C(0, 0, k, 1, 0)\rangle$ and $\langle cx_2, C(0, 0, k, 1, 1)\rangle$ have $m(\chi) = 2$.

References

[1] Frank W. Anderson and Kent R. Fuller (1973), *Rings and Categories of Modules*, Graduate Texts in Math., **13**. Springer-Verlag Inc., New York.

[2] Muhammed Salihu Audu (1983), "Transitive permutation groups of prime power order", PhD thesis, Queen's College, Oxford.

[3] T.R. Berger and L.G. Kovács and M.F. Newman (1980), "Groups of prime power order with cyclic Frattini subgroup", *Nederl. Akad. Wetensch. Proc. Ser. A*, **83**(1), 13–18.

[4] Kenneth S. Brown (1982), *Cohomology of Groups*, Graduate Texts in Math., **87**. Springer-Verlag, New York, Heidelberg, Berlin.

[5] S.B. Conlon (1976), "p-groups with an abelian maximal subgroup and cyclic center", *J. Austral. Math. Soc. Ser. A*, **22**, 221–233.

[6] C.W. Curtis and I. Reiner (1962), *Representation Theory of Finite Groups and Associative Algebras* (2nd edition), Pure and Appl. Math., **11**. Interscience, New York.

[7] Charles W. Curtis and Irving Reiner (1981), *Methods of Representation Theory (with Applications to Finite Groups and Orders), Volume 1.* John Wiley & Sons, New York, Chichester, Brisbane, Toronto.

[8] Walter Feit (1967), "Characters of Finite Groups", W.A. Benjamin, Inc., New York.

[9] Dane Laurence Flannery (1992), "Finite irreducible linear 2-groups of degree 4", PhD thesis, Australian National University.

[10] D.L. Flannery (1994), "Submodule lattices of direct sums", *Comm. Algebra*, **22**(10), 4067–4087.

[11] Marshall Hall, Jr. and James K. Senior (1964), *The Groups of Order* 2^n $(n \leq 6)$. Macmillan, New York.

[12] P.J. Hilton and U. Stammbach (1971), *A Course in Homological Algebra*, Graduate Texts in Math., **4**. Springer-Verlag, New York, Heidelberg, Berlin.

[13] I.M. Isaacs (1976), *Character Theory of Finite Groups*, Pure and Applied Mathematics, **69**. Academic Press, New York.

[14] C.R. Leedham-Green and W. Plesken (1986) "Some Remarks on Sylow Subgroups of General Linear Groups", *Math. Z.*, **191**, 529–535.

[15] Saunders Mac Lane (1963), *Homology*. Springer-Verlag, Berlin, Göttingen, Heidelberg.

[16] Derek J. Robinson (1981), "Applications of cohomology to the theory of groups", C.M. Campbell and E.F. Robertson (Eds.), *Group-St Andrews 1981*, LMS Lecture Notes, **71**, 46–80.

[17] Derek J. Robinson (1982), *A Course in the Theory of Groups*, Graduate Texts in Math., **80**. Springer-Verlag, New York, Heidelberg, Berlin.

[18] Edwin Weiss (1969), *Cohomology of Groups*, Pure and Applied Mathematics, **34**. Academic Press, New York.

Editorial Information

To be published in the *Memoirs*, a paper must be correct, new, nontrivial, and significant. Further, it must be well written and of interest to a substantial number of mathematicians. Piecemeal results, such as an inconclusive step toward an unproved major theorem or a minor variation on a known result, are in general not acceptable for publication. *Transactions* Editors shall solicit and encourage publication of worthy papers. Papers appearing in *Memoirs* are generally longer than those appearing in *Transactions* with which it shares an editorial committee.

As of May 31, 1997, the backlog for this journal was approximately 8 volumes. This estimate is the result of dividing the number of manuscripts for this journal in the Providence office that have not yet gone to the printer on the above date by the average number of monographs per volume over the previous twelve months, reduced by the number of issues published in four months (the time necessary for preparing an issue for the printer). (There are 6 volumes per year, each containing at least 4 numbers.)

A Copyright Transfer Agreement is required before a paper will be published in this journal. By submitting a paper to this journal, authors certify that the manuscript has not been submitted to nor is it under consideration for publication by another journal, conference proceedings, or similar publication.

Information for Authors and Editors

Memoirs are printed by photo-offset from camera copy fully prepared by the author. This means that the finished book will look exactly like the copy submitted.

The paper must contain a *descriptive title* and an *abstract* that summarizes the article in language suitable for workers in the general field (algebra, analysis, etc.). The *descriptive title* should be short, but informative; useless or vague phrases such as "some remarks about" or "concerning" should be avoided. The *abstract* should be at least one complete sentence, and at most 300 words. Included with the footnotes to the paper, there should be the 1991 *Mathematics Subject Classification* representing the primary and secondary subjects of the article. This may be followed by a list of *key words and phrases* describing the subject matter of the article and taken from it. A list of the numbers may be found in the annual index of *Mathematical Reviews*, published with the December issue starting in 1990, as well as from the electronic service e-MATH [**telnet e-MATH.ams.org** (or **telnet 130.44.1.100**). Login and password are **e-math**]. For journal abbreviations used in bibliographies, see the list of serials in the latest *Mathematical Reviews* annual index. When the manuscript is submitted, authors should supply the editor with electronic addresses if available. These will be printed after the postal address at the end of each article.

Electronically prepared papers. The AMS encourages submission of electronically prepared papers in $\mathcal{A}_{\mathcal{M}}\mathcal{S}$-TeX or $\mathcal{A}_{\mathcal{M}}\mathcal{S}$-LaTeX. The Society has prepared author packages for each AMS publication. Author packages include instructions for preparing electronic papers, the *AMS Author Handbook*, samples, and a style file that generates the particular design specifications of that publication series for both $\mathcal{A}_{\mathcal{M}}\mathcal{S}$-TeX and $\mathcal{A}_{\mathcal{M}}\mathcal{S}$-LaTeX.

Authors with FTP access may retrieve an author package from the Society's Internet node `e-MATH.ams.org` (130.44.1.100). For those without FTP access, the author package can be obtained free of charge by sending e-mail to `pub@math.ams.org` (Internet) or from the Publication Division, American Mathematical Society, P.O. Box 6248, Providence, RI 02940-6248. When requesting an author package, please specify $\mathcal{A}\mathcal{M}\mathcal{S}$-TEX or $\mathcal{A}\mathcal{M}\mathcal{S}$-LATEX, Macintosh or IBM (3.5) format, and the publication in which your paper will appear. Please be sure to include your complete mailing address.

Submission of electronic files. At the time of submission, the source file(s) should be sent to the Providence office (this includes any TEX source file, any graphics files, and the DVI or PostScript file).

Before sending the source file, be sure you have proofread your paper carefully. The files you send must be the EXACT files used to generate the proof copy that was accepted for publication. For all publications, authors are required to send a printed copy of their paper, which exactly matches the copy approved for publication, along with any graphics that will appear in the paper.

TEX files may be submitted by email, FTP, or on diskette. The DVI file(s) and PostScript files should be submitted only by FTP or on diskette unless they are encoded properly to submit through e-mail. (DVI files are binary and PostScript files tend to be very large.)

Files sent by electronic mail should be addressed to the Internet address `pub-submit@math.ams.org`. The subject line of the message should include the publication code to identify it as a Memoir. TEX source files, DVI files, and PostScript files can be transferred over the Internet by FTP to the Internet node `e-math.ams.org` (130.44.1.100).

Electronic graphics. Figures may be submitted to the AMS in an electronic format. The AMS recommends that graphics created electronically be saved in Encapsulated PostScript (EPS) format. This includes graphics originated via a graphics application as well as scanned photographs or other computer-generated images.

If the graphics package used does not support EPS output, the graphics file should be saved in one of the standard graphics formats—such as TIFF, PICT, GIF, etc.—rather than in an application-dependent format. Graphics files submitted in an application-dependent format are not likely to be used. No matter what method was used to produce the graphic, it is necessary to provide a paper copy to the AMS.

Authors using graphics packages for the creation of electronic art should also avoid the use of any lines thinner than 0.5 points in width. Many graphics packages allow the user to specify a "hairline" for a very thin line. Hairlines often look acceptable when proofed on a typical laser printer. However, when produced on a high-resolution laser imagesetter, hairlines become nearly invisible and will be lost entirely in the final printing process.

Screens should be set to values between 15% and 85%. Screens which fall outside of this range are too light or too dark to print correctly.

Any inquiries concerning a paper that has been accepted for publication should be sent directly to the Editorial Department, American Mathematical Society, P. O. Box 6248, Providence, RI 02940-6248.

Editors

Selected Titles in This Series

(*Continued from the front of this publication*)

(See the AMS catalog for earlier titles)